LICENSING GEOGRAPHIC DATA AND SERVICES

Committee on Licensing Geographic Data and Services
Board on Earth Sciences and Resources
Division on Earth and Life Studies

NATIONAL RESEARCH COUNCIL
OF THE NATIONAL ACADEMIES

THE NATIONAL ACADEMIES PRESS
Washington, D.C.
www.nap.edu

THE NATIONAL ACADEMIES PRESS 500 Fifth Street, N.W., Washington, DC 20001

NOTICE: The project that is the subject of this report was approved by the Governing Board of the National Research Council, whose members are drawn from the councils of the National Academy of Sciences, the National Academy of Engineering, and the Institute of Medicine. The members of the committee responsible for the report were chosen for their special competences and with regard for appropriate balance.

This study was supported by Department of Commerce/National Oceanic and Atmospheric Administration, Grant No. 50-DGNA-1-900024, Department of Commerce/U.S. Census Bureau Grant No. 43-YA-BC-154376, Department of Interior/U.S. Geological Survey Grant No. 02HQAG0150, Federal Emergency Management Agency Grant No. EMW-2002-SA-0210, and U.S. Government Printing Office Grant No. 733804. Any opinions, findings, conclusions, or recommendations expressed in this publication are those of the author(s) and do not necessarily reflect the views of the organizations or agencies that provided support for the project.

International Standard Book Number 0-309-09267-1 (Book)
International Standard Book Number 0-309-54457-2 (PDF)

Additional copies of this report are available from the National Academies Press, 500 Fifth Street, N.W., Lockbox 285, Washington, DC 20055; (800) 624-6242 or (202) 334-3313 (in the Washington metropolitan area); Internet, http://www.nap.edu

Cover design by Michele de la Menardiere; satellite images courtesy of NASA.

Printed in the United States of America

THE NATIONAL ACADEMIES

Advisers to the Nation on Science, Engineering, and Medicine

The **National Academy of Sciences** is a private, nonprofit, self-perpetuating society of distinguished scholars engaged in scientific and engineering research, dedicated to the furtherance of science and technology and to their use for the general welfare. Upon the authority of the charter granted to it by the Congress in 1863, the Academy has a mandate that requires it to advise the federal government on scientific and technical matters. Dr. Bruce M. Alberts is president of the National Academy of Sciences.

The **National Academy of Engineering** was established in 1964, under the charter of the National Academy of Sciences, as a parallel organization of outstanding engineers. It is autonomous in its administration and in the selection of its members, sharing with the National Academy of Sciences the responsibility for advising the federal government. The National Academy of Engineering also sponsors engineering programs aimed at meeting national needs, encourages education and research, and recognizes the superior achievements of engineers. Dr. Wm. A. Wulf is president of the National Academy of Engineering.

The **Institute of Medicine** was established in 1970 by the National Academy of Sciences to secure the services of eminent members of appropriate professions in the examination of policy matters pertaining to the health of the public. The Institute acts under the responsibility given to the National Academy of Sciences by its congressional charter to be an adviser to the federal government and, upon its own initiative, to identify issues of medical care, research, and education. Dr. Harvey V. Fineberg is president of the Institute of Medicine.

The **National Research Council** was organized by the National Academy of Sciences in 1916 to associate the broad community of science and technology with the Academy's purposes of furthering knowledge and advising the federal government. Functioning in accordance with general policies determined by the Academy, the Council has become the principal operating agency of both the National Academy of Sciences and the National Academy of Engineering in providing services to the government, the public, and the scientific and engineering communities. The Council is administered jointly by both Academies and the Institute of Medicine. Dr. Bruce M. Alberts and Dr. Wm. A. Wulf are chairman and vice chairman, respectively, of the National Research Council.

www.national-academies.org

LICENSING GEOGRAPHIC DATA AND SERVICES COMMITTEE

HARLAN J. ONSRUD, *Chair*, University of Maine, Orono
PRUDENCE S. ADLER, Association of Research Libraries, Washington, D.C.
HUGH N. ARCHER, Department for Natural Resources, Frankfort Kentucky
STANLEY M. BESEN, Charles River Associates, Washington, D.C.
JOHN W. FRAZIER, Binghamton University, Binghamton, New York
KATHLEEN (KASS) GREEN, Independent Consultant, Berkeley, California
WILLIAM S. HOLLAND, GeoAnalytics, Inc., Madison, Wisconsin
TERRENCE J. KEATING, Z/I Imaging, Madison, Alabama
JEFF LABONTÉ, GeoConnections Programs, Ottawa, Ontario, Canada
XAVIER R. LOPEZ, Oracle Corporation, Nashua, New Hampshire
STEPHEN M. MAURER, Private Practice/University of California, Berkeley
SUSAN R. POULTER, University of Utah, Salt Lake City
MARK E. REICHARDT, Open GIS Consortium, Inc., Gaithersburg, Maryland
TSERING W. SHAWA, Princeton University, Princeton, New Jersey

National Research Council Staff

PAUL M. CUTLER, Senior Program Officer
VERNA J. BOWEN, Financial and Administrative Associate
KAREN L. IMHOF, Senior Project Assistant (until December 2003)
MONICA LIPSCOMB, Research Assistant

MAPPING SCIENCE COMMITTEE

DAVID J. COWEN, *Chair*, University of South Carolina, Columbia
KATHRINE CARGO, Orleans Parish Communications District, New Orleans, Louisiana
KEITH C. CLARKE, University of California, Santa Barbara
WILLIAM J. CRAIG; University of Minnesota, Minneapolis
ISABEL F. CRUZ, University of Illinois at Chicago
ROBERT P. DENARO, Navteq Inc., Chicago, Illinois
SHOREH ELHAMI, Delaware County Auditor's Office, Delaware, Ohio
DAVID R. FLETCHER, Geographic Paradigm Computing, Albuquerque, New Mexico
HON. JAMES GERINGER, Environmental Systems Research Institute, Inc., Wheatland, Wyoming
DAVID R. MAIDMENT, University of Texas at Austin
MARK MONMONIER, Syracuse University, Syracuse, New York
JOEL MORRISON, Ohio State University (Retired), Reston, Virginia
SHASHI SHEKHAR, University of Minnesota, Minneapolis
NANCY TOSTA, Ross & Associates Environmental Consulting, Ltd., Seattle, Washington

National Research Council Staff

PAUL M. CUTLER, Senior Program Officer

BOARD ON EARTH SCIENCES AND RESOURCES

National Research Council Staff

Acknowledgments

This report has been reviewed in draft form by individuals chosen for their diverse perspectives and technical expertise, in accordance with procedures approved by the NRC's Report Review Committee. The purpose of this independent review is to provide candid and critical comments that will assist the institution in making its published report as sound as possible and to ensure that the report meets institutional standards for objectivity, evidence, and responsiveness to the study charge. The review comments and draft manuscript remain confidential to protect the integrity of the deliberative process. We wish to thank the following individuals for their review of this report:

John Copple, Sanborn Maps Company, Inc., Colorado Springs, Colorado
Shoreh Elhami, Delaware County Auditor's Office, Delaware, Ohio
Joanne Gabrynowicz, University of Mississippi, University
Guillermo Herrera, Bowdoin College, Maine. Woods Hole Oceanographic Institution, Woods Hole, MA
John R. Jensen, University of South Carolina, Columbia
Dennis Karjala, Arizona State University, Tempe
Donald T. Lauer, U.S. Geological Survey, retired, Santa Barbara, California
Robert Marx, U.S. Census Bureau, retired, Washington, D.C.
Roy Radner, New York University, New York City

Larry J. Sugerbaker, NatureServe, Arlington, Virginia

Although the reviewers listed above have provided many constructive comments and suggestions, they were not asked to endorse the conclusions or recommendations nor did they see the final draft of the report before its release. The review of this report was overseen by Shelton S. Alexander, Pennsylvania State University, University Park and Pierre C. Hohenberg, New York University, New York. Appointed by the National Research Council, they were responsible for making certain that an independent examination of this report was carried out in accordance with institutional procedures and that all review comments were carefully considered. Responsibility for the final content of this report rests entirely with the authoring committee and the institution.

Preface

Since 1988, the National Research Council's Mapping Science Committee (MSC) has provided independent advice to government and society on scientific, technical, and policy matters relating to geographic data.[1] The need for the present study was first expressed by federal agencies at a 2000 meeting of the MSC, and five agencies agreed to provide sponsorship: Federal Emergency Management Agency, Government Printing Office, National Oceanic and Atmospheric Administration, U.S. Census Bureau, and U.S. Geological Survey.

The study committee (Appendix A) held four information-gathering and writing meetings between February and November 2003, including a workshop that brought together approximately 60 people from government, academia, nongovernmental organizations, and the private sector. Presentations and white papers from the workshop are available at the MSC Web site. Along with requested written and verbal testimony from individuals (Appendix B), the committee analyzed published materials in developing its final report.

Because of the increasingly broad use of geographic information in society, and the growing use of licensing by data providers, the report's audience extends beyond the study's sponsors to government agencies at all levels, Congress, the private sector, academia, and the general public. The report is designed to be a resource for these groups. It distills the

[1]See <http://www7.nationalacademies.org/besr/Mapping_Science.html>.

legal, economic, and public interest underpinnings of data distribution approaches, lays out viewpoints and experiences of licensing from all sectors of the geographic data community, and proposes strategies that could advance data use and accessibility to the benefit of all stakeholders in geographic data.

Contents

APPENDIXES

Executive Summary

Licensing[1] *has both advantages and disadvantages, and is one of a number of procurement options available to government. Agencies considering data acquisition and distribution alternatives must weigh all options in the context of their mandates, missions, and user needs; government efficiency and accountability; the public interest; the effect of their decisions on private markets; and budgetary realities. The culture of licensing is evolving as the geographic data community experiments with this tool, and licenses are becoming more flexible. In this environment, society will be best served by agencies (1) sharing contract negotiation experiences and techniques; (2) frequently refreshing their understanding of data acquisition and dissemination options and user needs; (3) encouraging unambiguous, standardized, and automated licensing; (4) using licensing to improve coordination of data acquisitions; (5) enhancing government institutions that coordinate acquisitions; and (6) investigating options for building a National Commons and Marketplace in Geographic*

[1]*License* or *licensing* of geographic data or a geographic work means a transaction or arrangement (usually a contract, in which there is an exchange of value) in which the acquiring party (i.e., the licensee) obtains information with restrictions on the licensee's rights to use or transfer the information. *Geographic works* are works incorporating geographic data that have been collected, aggregated, manipulated, or transformed in some manner. *Geographic services* refers to the processes of obtaining, processing, or providing geographic data or geographic works. For a glossary of terms, see Appendix E.

Information. [2] *Progress on all items could be seen in the near term, although some aspects of items 5 and 6 would require more time and resources.*

BACKGROUND

The shift from sale of physical books, maps, and other intellectual works to the licensing of digital data, information, and affiliated services represents a significant change in the communication of knowledge. This shift is altering the balance between public and private interests in geographic works and data, a balance that has heretofore provided a level of creativity and innovation unequaled elsewhere in the world.

The balance that prevailed until recently was that the federal agencies generally paid for full rights in geographic data, including the right to publish it freely, permitting the data to enter the public domain and become readily accessible for follow-on uses. The prevalence of digital media as the platform for large databases, however, has made it feasible[3] for data providers to contemplate multiple licenses of geographic databases, even when the data are made available online to large numbers of potential users. Thus, vendors in some instances would prefer to license data to government agencies, with restrictions on redistribution and reuse of the data. In this environment, such data may not enter the public domain, and the cost of access may preclude otherwise beneficial uses, including the development of new products and services and informing public discourse. The fundamental issue addressed by this report, therefore, is under what circumstances and to what extent should agencies accept limitations on the further distribution or use of geographic data they acquire from private and other governmental vendors.

The number of uses of geographic data has expanded rapidly with the evolution of geographic information systems that manage geographic

---

[2] *Geographic information commons* means a system for making geographic data and works openly and freely accessible to the public over the Internet. A geographic information commons may include both public domain (i.e., free from any use restrictions) and open access content (i.e., content generally available for others to access, use, and copy, and often to make derivative works, although some limited restrictions may apply). *Geographic information marketplace* means a system for making geographic data and works available for sale over the Internet.

[3] *Feasibility* here refers to technological feasibility to limit access and in some instances to monitor or impede downstream uses, and to legal feasibility. The economic feasibility of this model will vary with the circumstances.

data,[4] improvements in remote-sensing technologies, the advent of inexpensive Global Positioning System receivers, the decreasing costs of personal computing and digital storage, the increasing reach of the Internet, and the increasing pervasiveness of wireless, location-aware telecommunications services. These developments have been accompanied by increased use of licensing as an alternative to the outright sale of the data and data products. Licensing has become commonplace because of

- the realization that many geographic data, as opposed to geographic creative works, are difficult to protect through copyright alone;
- a shift away from supplying distinct datasets to providing access to databases;
- the rise of business models that stress multiple subscribers despite the reality of digital networks and media that allow others to distribute perfect and inexpensive copies;
- increased concern over potential liability and a desire to limit liability through explicit license language; and
- the rise of shared cost and data maintenance partnerships.

Expanded mapping activities have increased the potential for duplication of data gathering and processing. Initiatives such as the U.S. Geological Survey's (USGS's) *The National Map*, the Office of Management and Budget's (OMB's) Geospatial One-Stop, and the U.S. Census Bureau's MAF/TIGER[5] modernization program seek to leverage local investments in geographic data and avoid unnecessary duplication. Because states, tribes, regional groups, counties, and cities have a wide range of data-sharing policies, the federal government is increasingly forced to address licensing issues for the data it acquires. Confusion and uncertainty have arisen as a result of

- a proliferation of nonstandard licensing arrangements;[6]
- difficulty in designing licenses that track the legal, economic, and public interest concerns of different levels of government;
- difficulty in designing licenses that accommodate all sectors of the geographic data community;

---

[4]See Appendix C for a description of the scope of geographic data.

[5]Master Address File/Topologically Integrated Geographic Encoding and Referencing system.

[6]See Appendix D for examples of the variety of licensing models.

- an imperfect appreciation for the licensing perspectives of different sectors of the geographic data community; and
- lack of effective license tracking and enforcement mechanisms.

THE COMMITTEE'S TASK

Given the confusion and uncertainty surrounding licensing, the National Academies, at the request of the Federal Emergency Management Agency, the Government Printing Office, the National Oceanic and Atmospheric Administration, the U.S. Census Bureau, and USGS, convened the Committee on Licensing of Geographic Data and Services in 2002. The committee was charged with six tasks:

1. to explore the experiences of federal, state, and local government agencies in licensing geographic data and services from and to the private sector using case studies such as the Landsat program;
2. to examine ways in which licensing of geographic data and services between government and the private sector serve agency missions and the interests of other stakeholders in government datasets;
3. to identify arguments in favor of and in opposition to spatial-data licensing arrangements;
4. to dissect newly proposed license-based models that could meet, concurrently, the spatial-data needs of government, the commercial sector, scientists, educators, and citizens;
5. to consider potential effects on spatial-data uses and spatial-technology developments of competing license/nonlicense approaches within the commercial sector, and
6. to analyze options that will balance the interests of all parties affected by licensing of spatial data[7] and services to and from government. Each of these tasks is now addressed in turn.

GOVERNMENT EXPERIENCES IN LICENSING GEOGRAPHIC DATA AND SERVICES FROM AND TO THE PRIVATE SECTOR

Despite recent interest in licensing, most federal agencies still prefer full ownership rights in the data that they acquire when this option is available. Their reasons vary from increased flexibility in the use of such

[7]Although the terms *geographic data* and *spatial data* are used interchangeably in the Statement of Task, we adopt the former throughout this report.

data to supporting agency and federal mandates relating to access, dissemination, duplication avoidance, waste avoidance, and saving money. Nonetheless, all federal agencies that provided input to the committee have acquired data under license from commercial vendors. Their reasons for doing so vary from being able to make maps faster and less expensively to having no alternative source of data meeting specific needs. Reactions to licensing differ from agency to agency, although there appears to be a general consensus that any cost advantage offered by acquiring licensed data must be weighed against constraints on current and possible future use and the interest in free exchange of information. In some cases, the coordination, negotiation, and administration costs associated with licensing are higher than those of other procurement methods.

Federal agencies almost always distribute geographic data at or below marginal cost of distribution. Since the 1990s, however, many state and local governments have experimented with distribution of data using licenses to generate revenue from their data.[8] Ten years later, many of these entities have concluded that fee programs (1) cannot recover a significant fraction of government data budgets, (2) seldom cover operating expenses, and (3) act as a drag on private-sector investments that would otherwise add to the tax base and grow the economy. The use of licenses to provide data to users may, however, be useful to enforce proper attribution, minimize liability, enhance data security, and formalize collaboration.

WAYS IN WHICH LICENSING SERVES AGENCY MISSIONS AND THE INTERESTS OF STAKEHOLDERS IN GOVERNMENT DATA

Agency mandates and missions can be broadly grouped into those requiring broad, limited, or internal data redistribution; those requiring distribution of derivative products; and those ensuring adequate citizen access and judicial review. In addition to utilizing outright purchases of data, agencies have experimented with a range of licenses to acquire data to perform their missions.[9] So far, results have been mixed. For the most part, agencies whose missions require broad dissemination find acquisition

---

[8]Examples included Hennepin County, Minnesota; the State of Maryland; and various European weather services. See also examples cited in Open Data Consortium, 2003, *10 Ways to Support Your GIS Without Selling Data*, available at <http://www.opendataconsortium.org>.

[9]See Chapter 4, Sections 4.2.2 and 4.2.3, for common types of licenses and nonlicense alternatives used by government.

of licensed data less useful than agencies that have small numbers of users or need licensed data as an input for making derivative products. Over time, some agencies have learned to negotiate new types of licenses that potentially offer better value to both the agency and commercial data suppliers.

Commercial data vendors have mixed attitudes toward licensing. In general, providers whose business models depend on adding value to data gathered from local, state, and federal agencies tend to oppose government data acquisition through licensing. Providers whose primary business models involve selling imagery or low-value-added geographic products to government generally welcome the prospect of licensing data to the government.

Academic users and producers are among the strongest advocates for the free flow of government geographic data as well as the free flow of any other publicly funded data and information of use to the scientific community. Nonetheless, the interests of students, teachers, researchers, libraries, and university administrators in gaining access to geographic data are not necessarily the same. For example, students and teachers may need legal and convenient access to data to accomplish class demonstrations, laboratory exercises, and class projects, but may care little about the right to openly publish datasets or derivative products. Researchers, on the other hand, need the legal and practical ability to access, use, and extend the datasets and work products of others, including the right to publish derivative works.

Ultimately, however, although agencies often are charged with promoting the public interest, the interests of actual and potential user groups may be discounted by agencies faced with budgetary constraints and vendors' demands.

ARGUMENTS IN FAVOR OF AND OPPOSITION TO LICENSING ARRANGEMENTS

The advantages of licensing data as opposed to outright purchase (e.g., through a for-hire service) include reducing acquisition costs (i.e., cost to the customer) in many instances, making data immediately available, enabling faster build times for operational information systems, structuring data release after a given embargo period, supporting specific agency projects as opposed to ongoing operations or decision-making functions, updating or correcting existing government databases, supporting national security uses, allocating risk, ensuring proper attribution, and supporting commercial markets.[10] Disadvantages of licensing can include increased acquisition cost in some instances; increased negotiation, coordination, administration, and enforcement costs; uncertain use and redistribution conditions; limited redistribution rights; inability to meet specialized needs; and loss of public domain effects. Restrictions on dissemination of data obtained by government from the commercial sector can impose large costs if these data are important inputs to research and development by businesses or the academic community. Nonetheless, licenses continue to evolve rapidly and are likely to improve over time. Suggestions from the commercial sector for promoting licenses include better contract design, validating licensed data to increase user confidence, developing standard form licenses, and simplifying negotiations.

LICENSE-BASED MODELS THAT COULD MEET THE GEOGRAPHIC DATA NEEDS OF ALL STAKEHOLDERS

The suitability of a particular license model depends on the intended use of the data. Furthermore, agencies must weigh the perspectives of many different stakeholders to find licenses that best fit their missions and mandates; legal, economic, and public interest considerations; and budgetary realities. Depending on the circumstances, the advantages of licensing may outweigh social and economic drawbacks of acquiring geographic data with some level of use restriction.

Legal Considerations[11]

Licensing of geographic data and works has come of age because of the limited protection afforded by copyright and other intellectual property doctrines in the digital environment. Providers also have turned to technological means to control access and copying, measures that are reinforced by the Digital Millennium Copyright Act for works that have at least some copyright protection. Moreover, courts have upheld contracts or licenses that limit the uses that a licensee can make of data, or that prohibit further distribution. Data providers' rights are likely to be further strengthened if Congress adopts database protection.

---

[10]See Chapter 4, Sections 4.2.5.1 and 4.2.5.2, for further explanation of each advantage and disadvantage listed here.

[11]See Chapter 5 for a more complete discussion.

Federal agency data acquisitions are also constrained by a variety of federal laws and regulations. Some federal laws and policies embody a strong preference for making data available to the public. Additionally, government accountability may require further public access, particularly in light of changes to the law regarding data access and data quality. Even so, documents such as OMB Circular No. A-130, the Federal Acquisition Regulations, and the Freedom of Information Act (FOIA) recognize the possibility that some government information will be subject to proprietary restrictions and cannot be disseminated or made public.

Economic Considerations[12]

Society makes data investment decisions through two very different institutions: governments and markets. Deciding which sector should acquire and distribute a particular product has profound implications for economic efficiency. For goods such as data, markets are a good solution where the initial decision to invest in a particular product is controversial or uncertain. Conversely, government procurement is most useful when uncertainty about whether to invest is small, so that distributional efficiency becomes the dominant concern. Beyond these generalizations, additional considerations may apply to particular cases.

Agencies affect the government/market balance each time they acquire or distribute data. The challenge is to make these choices with an eye toward economic efficiency. License design can be an important tool for setting this balance. In general, licenses for data obtained from the private sector that contain modest use and redistribution rights promote markets by allowing the original suppliers to pursue additional sales. Conversely, the agency may think that certain geographic data products have proven their worth, but that high prices are preventing many people from using them. In this case, the agency may wish to make the data it acquires widely available by acquiring them through licenses that give the agency broad redistribution rights.[13] Although such rights will limit any remaining private market for the product, that will be reflected in the price that the vendor demands, and the agency pays, for such a license. In return, efficiency in distribution is more likely to be achieved.

Traditional licensing models are not the only—or, in some cases, the best—ways to promote economic efficiency. For example, some firms will

---

[12]See Chapter 6 for a more complete discussion.

[13]In the limit, the agency may wish to purchase the data outright.

not mount large data acquisition programs unless agencies commit upfront resources. This can be done in a variety of ways, including private–public partnerships, licenses that obligate the agency to pay for large volumes of data, and cooperative research and development agreements. From the agency perspective, such transactions offer a good mix of efficiency in both production and distribution. Efficiency in production is achieved because the project must still realize significant private-sector sales to be profitable. Efficiency in distribution occurs because the agency often has significant leverage to demand license terms that permit widespread dissemination on favorable terms, perhaps by requiring the vendor to donate its data to the public after a fixed period of years, or otherwise limiting the private partner's ability to impose high prices. Alternatively, government might bear the entire cost of data production and acquire unlimited rights in order to promote efficiency in distribution.

The Public Interest[14]

Public discourse, equality, and innovation are benefits that are not easily assessed but accrue for society as a whole. These benefits have been well served by public domain[15] data, which have been the norm under a legal regime in which geographic data, once published, were free for anyone to use. Such data also serve government accountability and transparency, although some license restrictions also may support these public interests in some cases. National security, law enforcement, and privacy issues present a common challenge to policy makers considering geographic data access issues: how to weigh potentially harmful or intrusive uses against legitimate uses. Blanket restrictions and classification on national security or law enforcement grounds are inadvisable except in unambiguous cases. Furthermore, the potential benefits of classified data beyond the national security arena make timely declassification important. When classification is deemed appropriate, licenses can be used to limit access to specified users. Government also can use licenses to promote reuse of geographic data by negotiating terms that limit commercial firms' ability to discard data prematurely, promoting uniform and high-quality metadata,[16] and encouraging standards that make geographic

---

[14]See Chapter 7 for a more complete discussion.

[15]See Chapter 1, Section 1.4, for the definition of *public domain*.

[16]*Metadata* is information about data; for example, it might record such details as the collector, the sensor used, and when the data were collected (see

data interoperable across a wide range of hardware, software, and data products.

Recommendation 1: Before entering into data acquisition negotiations, agencies should confirm the extent of data redistribution required by their mandates and missions, government information policies, needs across government, and the public interest.[17]

Recommendation 2: Agencies should experiment with a wide variety of data procurement methods in order to maximize the excess of benefits over costs.[18]

Circumstances in Which the Need for Public Access Is Strong

When government uses data to promulgate regulations, formulate policy, or take other actions that affect the rights and obligations of citizens, there is a compelling interest in making these data available so that the public may understand, support, or challenge government decisions.[19] This interest often will be served by acquiring unlimited rights in data, but also may be accommodated in some circumstances by licensing data under conditions that permit access for more limited purposes. For example, in some cases the public may only need access to views derived from satellite data, rather than the original satellite data. The important principle is that access to information cannot be so limited, its distribution so difficult, or its content so closely held by government that outcomes of political debates are determined by unequal access to data.

Recommendation 3: When geographic data are to be used to design or administer regulatory schemes or formulate policy, affect the rights and obligations of citizens, or have likely value for the broader society as indicated by a legislative or regulatory mandate, the agency should

---

Federal Geographic Data Committee, 1998, *Data Content Standard for Digital Geospatial Metadata*, available at <http://www.fgdc.gov/standards/documents/standards/metadata/v2_0698.pdf>.)

[17]See Chapter 8, Section 8.3.

[18]See Chapter 8, Section 8.3.1.

[19]FOIA cannot be used to compel disclosure of legitimate trade secrets or proprietary information. However, an agency may not be able to support its decisions in court or elsewhere if the public or a court cannot scrutinize the evidence on which it relied.

evaluate whether the data should be acquired under terms that permit unlimited public access or whether more limited access may suffice to support the agency's mandates and missions and the agency's actions in judicial or other review.[20]

Strategies for Government Data Acquisition Under License

Compared to other procurement methods, the benefits and costs of licensing tend to be complex. The importance of particular terms usually depends on context. Thus, there is no "golden rule" for determining which license restrictions are appropriate. That said, agencies usually need to weigh such terms as price, dissemination restrictions, enforcement, available uplift rights,[21] and liability.

Recommendation 4: Agencies should agree to license restrictions only when doing so is consistent with their mandates, missions, and the user groups they serve.[22]

Although agencies are familiar with their own internal needs, it is important that they confirm that use restrictions are also acceptable to outside user groups included in their mandates and missions. This usually requires repeated, direct discussions with affected parties. Agencies also are trustees for the taxpayers they serve. This status includes an obligation to adhere to government's information policies by acquiring data that meet needs across government and serve the public interest. Because beneficial downstream uses and the public's interest in the free flow of information cannot be fully anticipated, agencies should exercise caution in construing their mandates and missions to permit licenses that restrict such uses.

Recommendation 5: Agencies that acquire data for redistribution should take affirmative steps to learn the needs and preferences of groups that are the intended beneficiaries of the data as defined by the mandates and missions of the agency. Agencies should avoid

---

[20]See Chapter 8, Section 8.3.2.3.

[21]*Uplift rights* in a license allow future purchases by specified parties under specified terms and conditions without the need to negotiate a new license.

[22]See Chapter 8, Section 8.3.2.4.

making technical choices in anticipation of secondary and tertiary uses[23] or consumer preferences.[24]

The Agency as Licensor

Most federal agencies and many state and local agencies routinely limit the fees they charge for data to the marginal cost of distribution.[25] However, federal agencies sometimes charge higher fees under specific statutory exceptions or to honor commercial restrictions on data previously obtained under license. Some state and local government agencies may choose (or are legally required) to set fees above marginal cost. In this case, agencies could limit the impact of such a decision by adopting the following strategies: (1) adopt price discrimination to mitigate the economic inefficiencies associated with user fees, (2) charge the lowest price consistent with covering their variable costs (if the goal is to finance ongoing operations while still providing affordable public access), and (3) present minimally restrictive contracts that offer a broad menu of licensing options. Even when cost recovery is not a goal, agencies sometimes may use licenses to pursue other, nonfinancial policy goals (e.g., ensuring attribution, negating implied endorsements, and managing risk). These provisions should impose minimal restrictions on licensees' ability to use and redistribute data.

Accommodating a Culture of Licensing

Most data vendors' standard terms and prices are negotiable, particularly for large transactions. When circumstances permit, agencies may want to demand fewer rights in exchange for lower prices. Agencies also

---

[23]*Secondary users* are those who are not the intended direct beneficiaries of the government data as defined by the mandates and missions of the agency but who nevertheless access government data and use it directly. *Tertiary users* are any users further downstream who do not directly acquire the data from government but gain access through others who pass it on with or without major changes.

[24]See Chapter 8, section 8.3.2.4.

[25]*Marginal cost* is the cost of providing a copy to an additional user. In some instances the cost of filling a user request requires additional preparation for the specific request, and the cost of that preparation along with the cost of duplication and delivery are included within the marginal cost.

may be able to offer in-kind payments to vendors in order to lower dollar costs still further.

Given the expansion of licensing of geographic data in the marketplace, agencies cannot help becoming more sophisticated consumers when licensing provides the best value or is the only available means of acquiring geographic data. The committee learned of many positive examples of agency negotiation. However, some agencies seemed to believe that they could not negotiate from a position of strength or found negotiations burdensome. As a result, some agencies indicated that they accepted vendors' opening offers at face value with little or no negotiation. Not coincidentally, these same agencies tended to have the most disappointing licensing experiences.

Recommendation 6: Agencies should dedicate resources to training and knowledge-sharing among agencies in order to extract maximum public benefit from licensing. The Federal Geographic Data Committee's working group and subcommittee structure provides a convenient venue through which agencies can report and learn from their experiences.[26]

POTENTIAL EFFECTS ON GEOGRAPHIC DATA USES AND GEOGRAPHIC TECHNOLOGY DEVELOPMENTS OF COMPETING LICENSE/NON-LICENSE APPROACHES WITHIN THE COMMERCIAL SECTOR

An earlier section of this summary[27] presented the contrasting interests of various groups in licensing and some of their concerns about effects of license and nonlicense approaches on geographic data use and technology. In addition, vignettes between the chapters of this report present visions for the effects of license and nonlicense approaches. Realization of these visions hinges on whether policy and/or technological solutions can be developed to address a license or nonlicense option. With each vignette, the vision builds in complexity to illustrate a possible future that accommodates the broadest range of stakeholders in geographic data and services. Chapter 9 lays out specific strategies and institutions that could help the geographic data community reach this goal through positive effects on geographic data uses and in directing technology developments.

---

[26]See Chapter 8, Section 8.5.

[27]See "Ways in Which Licensing Serves Agency Missions and the Interests of Stakeholders in Government Data." See also Chapter 4.

OPTIONS THAT WILL BALANCE THE INTERESTS OF ALL PARTIES AFFECTED BY LICENSING

Standard Licenses and Form Agreements

Today's geographic data contracts span a wide range of language and levels of complexity. At a minimum, it should be feasible to standardize provisions covering liability, indemnity, attribution, jurisdiction, and choice of law. Standard language and (eventually) standard form licenses are key to advancing many of the ideas in this report. Likely benefits include reduced negotiation costs, reduced uncertainty, improved market efficiency, and greater ability to automate transactions.

Recommendation 7: Agencies, trade associations, and public interest groups should exercise leadership in promoting standard clauses and form licenses throughout the geographic data community.[28]

Coordinating Government Acquisitions

Agencies often develop interagency approaches to prevent duplicate procurement of data. One strategy is to rely on a "lead agency" to purchase licenses on behalf of a larger group. Alternatively, agencies can purchase uplift rights when they acquire licensed data that are likely to be reused elsewhere in government.[29] Institutions such as data brokerages or automated business-to-government purchasing systems could strengthen this strategy.[30] However, these reforms may not be practical in the near term.

Recommendation 8: Agencies should continue to keep abreast of data brokerage and automated purchasing system developments that might help them coordinate data acquisitions from competing vendors.[31]

[28]See Chapter 9, Section 9.2.1.

[29]See Chapter 8, Section 8.4.2.2, for a discussion of uplift rights.

[30]*Data brokerages* enable users to search for previously licensed data. *Business-to-government* purchasing systems enable automated purchasing of standardized commercial products by government.

[31]See Chapter 9, Section 9.2.2.2.

Toward an Integrated National Commons and Marketplace in Geographic Information

Facilitating the sharing of and trade in data through the development of an efficient and user-friendly system, including a well-organized commons connecting users and contributors and an efficient market connecting buyers and sellers, would be a valuable endeavor.[32] Although no such online environment currently exists for geographic data, *The National Map,* Geospatial One-Stop, and the National Spatial Data Infrastructure provide first steps.

The National Commons

The overarching goal of the geographic information commons is to create a broad and continually growing set of freely usable (i.e., no monetary charge for use) geographic data and products at local scales similar in effect to the public domain datasets and works created by federal agencies. To succeed, the commons could provide easy, effective, and integrated mechanisms that

- enable any geographic dataset creator to construct a license that grants others permission to use his or her data,
- enable novice creators to quickly generate accurate and substantive standardized metadata for a geographic data file,
- enable data contributors to take advantage of form liability disclaimers,
- embed identifiers automatically in any commons dataset so that future users can link back to and recover detailed metadata and license conditions,
- allow for deeper search capabilities of geographic data and metadata than are currently available, and
- provide a long-term archive for commons geographic datasets.

Not all local governments, private citizens, or private companies will want to make any or all of their geographic datasets or products available in the public domain or in a commons licensing environment. Nevertheless,

[32]Vendors understand the value of a national market. One vendor told the committee that he would cut prices by three-fourths in a market that let him reach agency buyers (testimony from David DeLorme).

more people likely will make their data freely available when the integrated mechanisms described above become available.

Recommendation 9: The geographic data community should consider a National Commons in Geographic Information where individuals can post and acquire commons-licensed geographic data. The proposed facility would make it easier for geographic data creators (including local to federal agencies) to document, license, and deliver their datasets to a common shared pool, and also would help the broader community to find, acquire, and use such data. Participation would be voluntary.[33]

The Marketplace

The Internet has greatly enhanced the ability of commercial businesses, government, nonprofit organizations, and citizens to find comercial geographic data meeting their needs. A National Marketplace in Geographic Information would provide an online environment where any seller or licensor, no matter how small, could efficiently post its offerings in a searchable form using a menu of standard licenses and metadata reporting. Would-be purchasers could search through thousands of offerings to find the geographic data that meet their technical and license condition needs.

In the simplest implementation of the marketplace, purchasers would obtain the data directly from the vendor after "clicking through" to contact its server. Minimal investment could provide a combined license–metadata creation capability for sellers and search capability for consumers within a short time. In more advanced implementations, the seller or licensor might define detailed but standard license or sale conditions tied to seller-defined pricing formulas and participate in automated financial transactions and downloading of products. Buyers would be able to accomplish efficient comparison-shopping and buy or license desired geographic data within minutes of finding it. Seller's accounts could be automatically credited with funds from product sales, and sellers would be able to alter their geographic data offerings, descriptions, license conditions, and pricing formulas at any time.

Recommendation 10: The geographic data community should consider a National Marketplace in Geographic Information where

[33]See Chapter 9, Section 9.3.1.

individuals can offer and acquire commercial geographic data. The proposed facility would make it easier for the geographic data community to offer, find, acquire, and use existing geographic data under license. Participation would be voluntary.[34]

Encouraging Data Donations to the National Commons and Marketplace

A potential add-on to the basic commons and marketplace facility is a "timed donation strategy." To encourage donations, the following rule might be adopted: *Creators who post a data file for sale over the "marketplace component" must at the same time deposit a copy of the data file in escrow to the secured archives of the National Commons and Marketplace. Escrowed files become available after five years through a commons license selected by the creator at the time of deposit or, if no commons license is generated, enter the public domain.*[35]

This strategy is a natural extension to current USGS policies that use licensing to draw data into the public domain.[36] The benefits of such a strategy include (1) offering voluntary participation, (2) encouraging agency culture to become more sensitive to commercial concerns and foster greater coordination between private and public sectors, (3) improving data archiving, and (4) reinvigorating the public domain in geographic data.

Recommendation 11: The geographic data community should consider a system of "data donations" in which anyone who sells data using the National Marketplace in Geographic Information automatically agrees to donate their data to the commons after a commercially reasonable time, which we provisionally set at five years.[37]

---

[34]See Chapter 9, Section 9.3.2.

[35]Five years seems reasonable given the shelf life of most commercial products, but the definitive number should be based on a detailed study of the market.

[36]See USGS Policy 01-NMD001 (April 2001): Agencies should "convert licensed data to the public domain data by negotiating termination dates for license restrictions. The appropriate termination date may vary depending on the specific data type."

[37]See Chapter 9, Section 9.4.1.

Operating the National Commons and Marketplace

Because they require similar software and hardware, the commons and marketplace components could be built simultaneously as a unified or closely integrated facility. Assuming consistent standards and processes, separate entities conceivably could host and operate the different components of the system. Whatever the chosen path, strong agency leadership will be needed to ensure that maximum benefits are achieved.

Recommendation 12: Federal agencies should investigate options for and encourage development of a National Commons and Marketplace in Geographic Information.[38]

[38]See Chapter 9, Section 9.5.

1

Introduction

1.1 BACKGROUND

The shift from sale of physical books, maps, and other intellectual works to the licensing of digital data, information, and affiliated services represents a significant change in the communication of knowledge. This shift is altering the balance between public and private interests in geographic works and data.

This study focuses on licensing of geographic data and services to and from government. The number of uses of geographic data has expanded rapidly with the evolution of geographic information systems that manage geographic data,[1] improvements in remote-sensing technologies, the advent of inexpensive Global Positioning System (GPS) receivers, decreasing costs of personal computing and digital storage, the increasing reach of the Internet, and the increasing pervasiveness of wireless, location-aware telecommunications services. These developments have been accompanied by increased use of licensing as an alternative to the outright sale of the data and data products. Licensing has become commonplace because of

- the realization that many geographic data, as opposed to geographic creative works, are difficult to protect through copyright alone;

[1] See Appendix C for a description of the scope of geographic data.

- a shift away from supplying distinct datasets to providing access to databases;
- the rise of business models that stress multiple subscribers despite the reality of digital networks and media that allow others to distribute perfect and inexpensive copies;
- increased concern over potential liability and a desire to limit liability through explicit license language; and
- the rise of shared cost and data maintenance partnerships.

Expanded mapping activities have increased the potential for duplication. Initiatives such as the U.S. Geological Survey's *National Map*, the Office of Management and Budget's Geospatial One-Stop, and the U.S. Census Bureau's MAF/TIGER program modernization[2] seek to leverage local government investments in geographic data and avoid unnecessary duplication. Because states, tribes, regional groups, counties, and cities have a wide range of data-sharing policies that includes sharing under license, the federal government is increasingly forced to address licensing issues. Confusion and uncertainty have arisen as a result of

- a proliferation of nonstandard licensing arrangements;
- difficulty in designing licenses that track the legal, economic, and public interest concerns of different levels of government;
- difficulty in designing licenses that accommodate all sectors of the geographic data community;
- an imperfect appreciation for the licensing perspectives of different sectors of the geographic data community; and
- lack of effective license tracking and enforcement mechanisms.

Even within a single sector, there can be multiple perspectives on licensing. Commercial firms that wish to supply data to the government typically want high prices and significant restrictions on reuse. On the other hand, commercial firms that wish to acquire data from the government usually want low prices and few if any restrictions on reuse. These competing interests within the commercial sector add to the confusion and uncertainty surrounding licensing.

Licensing is one among several social tools for pursuing economic and policy objectives. Commercial data providers typically use licenses

[2]For *The National Map*, see < http://nationalmap.usgs.gov/>; for Geospatial One-Stop, see < http://www.geo-one-stop.gov/>; for MAF/TIGER (Master Address File/Topologically Integrated Geographic Encoding and Referencing system) modernization, see <http://www.census.gov/geo/mod/overview.pdf>.

to protect and receive a return on investments. Some government data providers similarly view licensing from government as an opportunity to earn revenue. More commonly, however, they use licensing to effect such policy objectives as ensuring data integrity by developing relations with known parties to whom notices of corrections and limitations may be delivered, ensuring that the most current government data are used by individuals and businesses, enforcing credit and attribution, and organizing collaboration.

From the perspective of government agencies, ownership and licensing each have benefits and drawbacks. Ownership (i.e., unrestricted transfers or purchases of data) lets government offer citizens and the commercial sector broad open access to data and any public records derived from them. This enhances the ability of citizens to check on the functioning of government, lets individuals and businesses develop markets based on the use of government information, and promotes research and society's general education. In contrast, acquisition under license may restrict government's ability to disseminate the data it uses and derivative products it produces. In addition, the new burdens imposed by the need to administer licenses can add to government's overhead costs. Licenses, similarly, can add to transaction costs that commercial and nonprofit users incur to acquire government data.

Conversely, licensing can help agencies accomplish their missions more efficiently and cost-effectively. In many cases, it may be cheaper to acquire data under license than through outright purchase. Agencies may also be able to discontinue some data collection and processing tasks if accurate, reliable, and cost-effective data can be licensed from the private sector. Finally, assuming that the public's interest in the free flow of information is accommodated, licensing may allow government agencies to shift costs from taxpayers to users by charging fees for agency data and services, although some efficiencies may be lost if costs are shifted to users.

Designing a licensing policy that balances the needs of government agencies, the commercial sector, and private citizens and citizen groups requires detailed consideration of multiple legal, policy, regulatory, and technology issues. The committee provides this guidance using the categories laid out in its Statement of Task.

1.2 STATEMENT OF TASK

Given the climate of confusion involving licensing of geographic data and services, the committee was charged with six tasks:

1. Explore the experiences of federal, state, and local government agencies in licensing geographic data and services from and to the private sector, using case studies such as the Landsat Program.
2. Examine ways in which licensing of geographic data and services between government and the private sector serve agency missions and the interests of other stakeholders in government datasets.
3. Identify arguments in favor of and in opposition to spatial-data licensing arrangements.
4. Dissect newly proposed license-based models that could meet, concurrently, the spatial-data needs of government, the commercial sector, scientists, educators, and citizens.
5. Consider potential effects on spatial-data uses and spatial-technology developments of competing license/nonlicense approaches within the commercial sector.
6. Analyze options that will balance the interests of all parties affected by licensing of spatial data and services to and from government.

The report provides a self-contained roadmap to the foregoing issues by (1) capturing the range of arguments and experiences found in different sectors of the geographic data community, (2) providing an overview of related disciplines including copyright law and information economics, and (3) pointing the reader to more specialized resources when necessary. Although the report focuses on the needs of the civilian federal agencies that sponsored the study, other government, commercial, and nonprofit sectors are also considered. For this reason, the information and conclusions presented in this report apply to a broad range of agency missions, goals, policies, and legal constraints. The report is likely to be of interest to federal, state, and local agencies; commercial firms; academic professionals; and citizens alike.

1.3 REPORT STRUCTURE

The modular structure of this report lets readers with varied backgrounds and interests select the particular discussions that interest them.

Chapters 2 and 3 are background chapters: Chapter 2 explores societal goals that motivate government missions and data policies; Chapter 3 describes types of geographic data, interrelationships among market players, the value chain of geographic data, and exchange mechanisms within the geographic data marketplace.

Chapter 4 addresses items 1, 2, and 3 in the committee's Statement of Task: licensing experiences of stakeholder groups, ways in which licensing serves agency missions and interests of other stakeholders, and advantages and disadvantages of licensing as seen from different perspectives.

Chapters 5 through 7 present the legal, economic, and public interest arguments that relate to licensing geographic data and services.

Chapter 8 presents guidelines for deciding when licensing to and from government may be appropriate and, if so, under what terms. The chapter also addresses Task 4 by presenting license-based models that could satisfy the range of stakeholders.

Chapter 9, describes strategies that could make current licensing institutions more efficient and also more responsive to the interests of all affected parties (Task 6). The chapter also addresses aspects of Tasks 4 and 5 (downstream impacts of licensing).

The committee's recommendations, initially presented in Chapters 8 and 9, are collected together in Chapter 10.

A series of vignettes, or "dream scenarios" is dispersed among each of the chapters. Realization of these dreams hinges on whether policy and/or technological solutions can be developed to address a license or nonlicense option. With each vignette, the vision builds in complexity to illustrate a possible future that accommodates the broadest range of stakeholders in geographic data and services. Chapter 9 lays out specific strategies and institutions that can or could help the geographic data community reach this goal.

Lastly, the appendixes provide a range of resources. Appendix C contains background information on the scope of geographic information. Appendix D summarizes current licensing models. Appendix E contains a glossary of terms,[3] and Appendix F lists acronyms used in the report.

1.4 KEY TERMS

This report repeatedly uses such terms as "data," "information," "works," "services," "purchase," "license," "ownership," "public domain," "open access content," and "information commons." In the interests of clarity, we define them now, at the outset. Other authors sometimes use different definitions and the reader should keep these definitions clearly in mind to avoid misunderstandings.

---

[3]Some key terms are also described in Section 1.4 of this chapter.

Data and *information* are elements of an ascending hierarchy:[4]

- *Data* are facts and other raw material that may be processed into useful information.
- *Information* is data processed and rendered useful.
- *Knowledge* is information transformed into meaning through action of the human mind, such that it can be recorded and transmitted.
- *Understanding* is knowledge integrated with a world view and a personal perspective, existing entirely within the human mind.
- *Wisdom* is understanding made whole and generative within the human mind.

Works of *knowledge* such as books, journals, and maps provide context and convey meaning and are often protected by intellectual property law. Because they exist only in the human mind, *understanding* and *wisdom* are not recordable on other media.

Geographic data are any location-based data or facts that result from observation or measurement, or are acquired by standard mechanical, electronic, optical, or other sensors.

Geographic works are works incorporating geographic data that have been collected, aggregated, manipulated, or transformed in some manner. Examples include datasets and databases, and other products derived from geographic data, including but not limited to maps, models, and other visualizations involving geographic data.

Geographic information means either geographic data or works without distinction, and may encompass, but is not limited to, (1) location-based measurements and observations obtained through human cognition or through such technologies as satellite remote sensing, aerial photography, GPS, and mobile technologies; and (2) location-based information transformed as images, photographs, maps, models, and other visualizetions. Geographic data and works are not strictly location based but may also include, for example, spatial relationships, descriptions or attributes of geographic features, metadata, and additional types of information that are arranged, categorized, or accessed in reference to their geographic or

[4]W. Crawford and M. Gorham, 1995, *Future Libraries: Dreams, Madness and Reality*, American Library Association, Chicago, p. 5.

spatial location. Such information typically is presented in digital form and may be contained in databases.

Geographic services refer to the processes of obtaining, processing, or providing geographic data or geographic works. As used in this volume, the term refers to the provision of access to and use of preexisting data or databases, such as subscription to a particular online geo-based processing capability or subscription to a database allowing downloads when desired. In some contexts, the term "services" may connote geographic data or works provided for a single client, according to that client's specifications.

Purchase of geographic data refers to a transaction or arrangement (usually a contract, in which there is an exchange of value) in which the purchaser of the geographic data (which may be contained in a geographic work) obtains *unlimited* rights to use, copy, and disseminate the geographic data. From the standpoint of the provider of the information, such a transaction is a *sale of geographic data*. The provider may retain a copy of the information as well as the right to enter into other sales or licenses of the information. Such a transaction is equivalent to a license for unrestricted use of information (see definition of *license* below). The purchase or sale of geographic data should be distinguished from the purchase or sale of a *copy* of a geographic work, such as a map or a copyrighted database, which does *not* relinquish the seller's intellectual property rights unless those rights are expressly or impliedly conveyed.

License or *licensing* of geographic data or a geographic work means a transaction or arrangement (usually a contract, in which there is an exchange of value) in which the acquiring party (i.e., the licensee) obtains information with restrictions on the licensee's rights to use or transfer the information. Examples of such restrictions include limits on the persons or entities (such as other agencies or the public) to whom the information may be disclosed, limits on the purposes for which the information may be used, limits on the duration of the license, and provisions regarding ownership and use of products developed through the use of the licensed information.

Ownership of geographic data is an inherently ambiguous, though widely used, term. One theory of ownership is that what rights the law gives an owner of information should be determined solely by how much of an incentive is needed to ensure the optimal production and distribution of the type of information in question. If the law gives more protection to information than is needed to induce someone to create it and to make it available, this results in suboptimal availability and a deadweight loss to society. In this report, *ownership of geographic data or works* as applied to the licensor–licensee relationship means:

- With reference to a vendor or licensor, the owner is in possession of information that is not publicly known and holds the information as a trade secret. In the case of information to which copyright applies, the licensor is the owner of the copyright.
- With reference to a licensee, the licensee has possession of a copy of the information and has exclusive or nonexclusive rights to use and make the information available to others without restriction.

The meaning of the term *public domain* requires more explanation than a simple definition. The term has been described as "a question-begging concept," the meaning of which must be understood in context.[5] To provide that context, several issues must be considered.

First, public domain information is usually defined as information that is not protected by copyright or patent law,[6] and we include this concept in our definition. Moreover, for information to be part of the public domain, it must be available to the public; hence trade secrets are not part of the public domain, even when not protected by copyright or patent.[7]

This definition is inadequate for our purposes, however, because the use of digital media permits data providers to impose license or contract terms with limitations on the use or redistribution of data that have not been possible for information published in paper media.[8] In such instances, these contract or license rights have the same effects as traditional intellectual property rights, such as patent or copyright. It therefore seems

[5] *Mine Safety Appliances Co. v. Electric Storage Battery Co.,* 405 F.2d 901, 902 (C.C.P.A. 1969). See also J. Boyle, 2003, Foreword: The opposite of property? 66 *Law & Contemporary Problems* 1 (describing various authors' use of the term in the Duke Conference on the Public Domain).

[6] Patent law in any event does not protect information as such, but rather applications of information. Thus, information in patents is usually treated as part of the public domain. *See* 1-1 Milgrim on Trade Secrets § 1.06 (2003).

[7] See *Kewanee Oil Co. v. Bicron Corp.*, 416 U.S. 470 (1974).

[8] See discussion of *ProCD v. Zeidenberg,* 86 F.3d 1447 (7th Cir. 1996), regarding validity of such licenses.

appropriate to add a requirement of unrestricted public use to the definition of public domain.

Last, we considered whether the definition should address the issue of cost of access, since costly access could greatly curtail if not prohibit the use of nominally public domain data. In a robust competitive market, a data provider will be unable to price data above the cost of reproduction. Anyone who pays the price of access could establish a competing, lower price service due to lack of need to gather or produce the data in the first place.[9] Assuming the existence of well-functioning competitive markets, such data will be available at reasonable cost.

Considering all of the foregoing, it seems appropriate to define *public domain information* for the purposes of this report as information that is not protected by patent, copyright, or any other legal right, and is accessible to the public without contractual restrictions on redistribution or use.

Open access content, for purposes of this report, is content openly available for others to access, use, and copy, and often to make derivative works, although some limited restrictions may apply. Typical restrictions may include preventing users from removing creator attribution from content, imposing identical license terms on any derived works, barring commercial use without permission, and liability limitations. We note that this definition does not necessarily conform to the use of the phrase "open access" in other contexts, including scientific publishing.

Geographic information commons means a system for making geographic data and works openly and freely accessible to the public over the Internet. A geographic information commons may include both public domain (i.e., free from any use restrictions) and open access content.

Geographic information marketplace means a system for making geographic data and works available for sale over the Internet (see Chapter 9, Section 9.3).

[9] In fact, the price charged for access to the data should tend toward the marginal cost of distribution under accepted economic theory (Chapter 6, Section 6.2), though in practice this condition may not be reached.

VIGNETTE A. A TEACHER'S DREAM[10]

Mr. Henson is a high school science teacher, but his students call him a "tree freak." Each spring, he sends his students out with global positioning system receivers to locate and identify trees in the surrounding community. In the field, each student collects leaf samples from at least 50 tree species, records the location of each identified tree, and describes the bark, canopy shape, and evidence of pests and diseases. To serve as a base map for their project, the class downloaded a digital map from the Internet showing all streets and land parcels in the town. Ten years later, Henson has a database and leaf samples for several thousand trees. Hundreds of the trees have been observed multiple times over the years.

One day the local paper writes a story about Henson's class. Within hours, a local government official asks for a copy of the database. He wants to see whether a smelter built four years earlier may be affecting vegetation in the community. Two graduate students from a local university also ask to see Henson's data. They want to study how sudden oak death syndrome propagates in an urban environment. Henson is delighted by the attention and posts his database on the school's Web site. That afternoon he tells students that their work is being used in the real world. He even asks them to imagine how biologists, planners, and historians could still be using their data a century from now.

There's a problem, however. The next morning the school board's lawyer points out that Henson's class has used a base map of unknown source. She doesn't know whether the school can copy or redistribute it. The lawyer tries to contact a company that may have been the owner of the digital map but it has moved in the interim. A dejected Henson takes down the Web site and tells people he can't share his data until the legal issues are resolved or he can find an appropriate substitute digital base map.

Mr. Henson ponders. Might not a system be developed through which the source of all local maps and geographic data on the Web could be readily determined? Alternatively, how might one readily determine if digital maps are in the public domain?

[10]This and all other vignettes are included for illustrative purposes. They are designed to clarify concepts in the report.

2

Society's Goals

2.1 INTRODUCTION

Society's goals are reflected in government's evolving laws, policies, and institutions. Like other powerful, flexible technologies, spatial technologies and affiliated geographic data pose multiple opportunities and challenges. For this reason, geographic data policies that focus too narrowly on a single societal goal or issue are likely to have unintended consequences. Solutions need to balance *all* relevant goals, and agencies should identify these goals early and keep them constantly in mind as they weigh data policy decisions.

This chapter summarizes societal goals that agencies often pursue within their legislated missions and U.S. government information policy generally. Although the discussion focuses on federal agency goals, similar considerations apply to state and local agencies. We do not analyze what government might do to advance any particular goal, or how licensing might fit into such a strategy; that analysis is presented in Chapter 7.

2.2 PROMOTING GOVERNMENT ACCOUNTABILITY AND TRANSPARENCY

Democracy depends on government accountability and transparency. In part, this rests on access to government information. Without information, individuals cannot effectively participate in matters in which government affects their daily lives.

The Freedom of Information Act creates a balance between the rights of citizens to be informed about government activities and the need to maintain confidentiality of some government records. In many cases, political transparency may require distributing geographic data to anyone who wants it. Citizens need access to geographic data to become educated in the detailed functioning of government; to petition government agencies, lobby legislators, analyze regulatory decisions; or to challenge illegal actions and government abuses in court. Government uses geographic data to make myriad decisions, and citizens often cannot know whether inappropriate manipulation of data has occurred without access to the entire record. An important principle of democracy is that access to government information is a matter of equal protection—that is, all citizens should have the same rights to public information to understand and be able to challenge government actions.

2.3 MAXIMIZING NET BENEFITS: THE DIFFERENCE BETWEEN BENEFITS AND COSTS

Taxpayers have an interest in seeing that government maximizes the difference between benefits and costs when it performs its missions. Licensing may sometimes be the best way to achieve this goal, depending on costs and government's need to redistribute the data. Government's redistribution needs range from internal use to broad redistribution.

Costs are not limited to the license fees that government must pay. The concept of cost also extends beyond dollar royalties. In the case of licenses, it includes the costs of negotiating transactions, administering intellectual property rights management obligations, and enforcement in the event of disputes. Agencies must also acquire data of sufficient quality and quantity to perform their missions. To some extent, this requirement can be defined in such technical terms as geographic coverage, timeliness, frequency of updates, spatial resolution, and accuracy of annotations. Agencies also need sufficient use and redistribution rights to meet known needs and unexpected needs that might not evolve until much later. A cost-benefit analysis must consider both time frames.

A further consideration for agencies is that society may not obtain full value for its investment unless existing geographic data are used again and again. The benefits of data reuse have long been recognized and are reflected in various archiving strategies, ranging from traditional paper map libraries to today's electronic geolibraries.[1]

That being said, government's outright acquisition of data without license restrictions is no guarantee that use and reuse will exceed that of potential acquisition alternatives. In part, this is because government is primarily focused on governance, not data distribution.[2] Agencies need to resist such a parochial viewpoint. A comprehensive data strategy must ensure that the taxpayers' investment in data generates maximum value not just for government, but for the entire society.

Lastly, agencies can avoid purchasing duplicative or unnecessary data by broadening their view beyond isolated, one-off transactions. This is a particular challenge for the federal system, in which purchasing decisions tend to be dispersed among multiple agencies. This challenge also transcends licensing and exists even where data are purchased outright.

Pursuit of the public good demands deep inquiry in support of decision making. Government needs to take into account both short- and long-term impacts of its decisions when weighing benefits and costs.

2.4 OBTAINING DATA ON BEHALF OF SOCIETY

Society acquires and distributes data for many purposes. In some cases, this work is done by private enterprise. In others, government agencies collect or purchase data to make them available to the broader society. Since the early nineteenth century, the federal government has mounted massive mapping programs to support settlement, commerce, and exploitation and preservation of the nation's natural resources. Among the products of these programs are topographic maps, marine and aeronautical navigation charts, census data, and digital orthoimages.[3] In the twenty-first

---

[1]See National Research Council, 1999, *Distributed Geolibraries: Spatial Information Resources*, Washington, D.C., National Academies Press.

[2]Although there are many government data distribution mechanisms that aim to meet consumer demand and improve data reuse (e.g., Geospatial One-Stop, *The National Map*, and many local and state efforts), there are also numerous businesses that provide value-added services and make unrestricted government data more accessible (e.g., Topozone [by Maps a la carte, Inc.], Land Info [by LAND INFO International, LLC], and Maptech [by Maptech, Inc.]).

[3]An orthoimage is a specially processed image prepared from an aerial photograph or a remotely sensed image that combines the accuracy of a traditional line map with the detail of an aerial image.

century, government data acquisitions can similarly support development of basic resources for the information economy by laying the groundwork for new and more valuable products.

Agency missions can require acquisition and distribution of data, or both. Agencies can *acquire* geographic data by (1) having employees collect it, (2) hiring outside contractors to collect it, (3) purchasing preexisting data from the private sector, or (4) obtaining a license to use preexisting or newly collected data. Unlike the first three options, licensing does not give government unlimited rights to use and redistribute the data. Regardless of whether a federal agency purchases or licenses data from a commercial vendor,[4] the vendor typically remains free to license or sell the data to others. Licensing may be a useful way for agencies to acquire data if the existence of a private market reduces the price paid by taxpayers. In the case of *distribution*, agencies can pass data to users directly or through commercial intermediaries. Both options may—but need not—include licenses.

2.5 SUPPORTING GOVERNMENT MISSIONS AND INDIVIDUAL RIGHTS

Geographic data are widely used within and outside government to assist economic development, protect property rights, support education, maintain the nation's physical infrastructure, protect the environment, develop natural resources, support health care, protect national security, facilitate taxation, and ensure the safety, health, security, property, and privacy of individual citizens. For example, national security and law enforcement agencies may use anything from street-centerline data that provide a reference framework for tracking patrol cars to images of battlefields from space. In many cases, government officials require access to assets that also support civilian applications. Whether the asset is a readily duplicated information good (e.g., street-centerline data files) or a scarce physical asset (e.g., a remote-sensing satellite) often bears on the decision of how data will be procured, in particular whether licensing is a feasible option. For example, in the former case, there likely are alternative options in the marketplace, some of which may offer a greater difference between benefits and costs than licensing. In the case of satellite data,

---

[4]Recall from Chapter 1, Section 1.4, that we define purchase to mean the acquisition of unlimited rights, not the elimination of the ability of the seller to sell or license the data to other users.

there likely are fewer alternative sources and thus a greater likelihood that licensing represents a plausible alternative based on cost.

Governments and commercial businesses collect sensitive geographic information in support of their missions. National security and law enforcement agencies fear misuse of these data by hostile nations, terrorists, and criminals. On the other hand, businesses and government agencies at all levels need the data to make their operations more efficient and effective. Government must balance security concerns against legitimate uses and citizens' basic "right to know." In some cases, this may mean controlling access to information so that it reaches some people but not others. Nonetheless, widespread access to geographic data gathered by government, including declassified data, can have economic benefits. A strong economy is critical to national security.

Citizens' rights, including privacy, are profoundly affected by the collection and maintenance of geographic data by the commercial sector and government. Individuals, as well as corporations in some cases, have a privacy interest in controlling access to data that describe them. To some extent, the interest is based on fears that governments, corporations, or other individuals will misuse such information. Examples include files containing information on wealth, income, purchases, daily travel routes, or health; and satellite, aerial, and street-level images of private property. Society also protects some information even when misuse is not an issue. This is based on case law or legislative judgments that individuals should be allowed to control inherently "private" information.

2.6 SUMMARY

Government geographic data acquisition and dissemination policies must balance multiple social goals, including (1) promoting government accountability and transparency, (2) maximizing net benefits (the difference between benefits and costs), (3) obtaining data on behalf of society, and (4) supporting government missions and individual rights. Agencies must make these decisions in accordance with existing laws, regulations, government policies, and budget constraints. There are times when it is prudent for government to be a licensee or licensor of geographic data and times when it is not.

VIGNETTE B. A LOCAL GOVERNMENT'S DREAM

The staff at Franklin County's Department of Natural Resources want to create a digital map of soil erosion potential to help them advise on land development activities in sensitive areas. Because the watersheds that affect erosion extend well beyond the county boundary, Ed Johnson goes to an online portal to find and download the data he needs. These data include up-to-date, detailed aerial imagery of the watersheds from a commercial database; soil type data from the U.S. Department of Agriculture; vegetation coverage from the state department of natural resources, and the locations and sizes of culverts, bridges, channels, and other stormwater facilities from Franklin County and neighboring counties.

Mr. Johnson is able to acquire most of this information without negotiating use terms or paying substantial fees. Many local, state, and federal agencies, as well as a number of private parties, have placed some of their data into an online commons by using licenses that minimally constrain the downstream uses of the data. Additionally, the portal offers convenient "one-stop shopping" for many suppliers' data and reduces the cost of searching for data. Finally, standard online license forms reduce the complexity of licensing and the need for separate negotiations between Mr. Johnson and different data suppliers. This streamlining has brought down transaction costs.

Although several commercial aerial imagery offerings meet Mr. Johnson's technical requirements, he quickly selects and purchases the one set of commercial imagery offering the best combination of quality, price, and use rights for his needs.

In the end, the dream comes down to this: Can a Web portal based on standardized licensing be developed that efficiently supports an active information commons and a thriving marketplace in geographic data and services?

3

The Geographic Data Market: Offerings, Players, and Methods of Exchange

3.1 INTRODUCTION

Efforts to enhance geographic data production, distribution, and use—the underlying goal of government licensing policy—benefit from understanding the geographic data market. The interaction between public and private sectors in this market is complex. It is also changing, with vendors announcing new business models every few months.

This chapter begins by summarizing the types of geographic data that society uses. Next, it describes the structure of the marketplace and reviews the "value chain" that results from successive actors collecting, merging, and transforming raw data into products and services within the marketplace. The chapter then discusses dominant business models in geographic data transactions and the factors that influence contractual terms. The final sections summarize common license models and data flows to and from the public sector.

3.2 TYPES OF GEOGRAPHIC OFFERINGS

Today's digital technology permits the rapid acquisition and maintenance of a vast inventory of information about Earth, ranging from demographic descriptors to well-defined uses of land. Geographic data types can be categorized in many ways. Here, we differentiate data that describe the natural world from those describing human action.[1] Numerous natural features characterize Earth's physical processes, patterns, and conditions. These features, described in more detail in Appendix C, relate to topography, hydrology, physical geology and geography, weather, energy resources, and natural resources and hazards. In contrast, the constructed environment constitutes the human geography at Earth's surface. These features include built structures and invisible boundaries that reflect political, economic, and locational decisions. Such data can be broadly grouped into five categories of human action: (1) transportation, (2) institutional locations (e.g., colleges or universities, schools, and libraries; hospitals and nursing homes; industrial sites; parks and historic landmarks and sites), (3) energy-related infrastructure, (4) administrative and legal boundaries, and (5) hazardous locations.

3.2.1 Focus of Government Geographic Data Interests

Government agencies usually focus on data needed to address their own mandates, missions, and goals. Nonetheless, government information policy aims to ensure that most data gathered for one government purpose are widely available to support additional governmental and nongovernmental uses. Geographic data priorities vary by agency and level of government, but some data types are particularly versatile and tend to support multiple missions. In the federal government, the National Research Council (NRC)[2] identified three geographic data themes as being at the foundation of government business: Terrain (elevation) data, orthoimagery, and geodetic control.[3] NRC (1995) also highlighted additional "framework" data types that are often high priorities for agencies: transportation networks; political, administrative, and census boundaries; hydrology (location, geometry, and flow characteristics of rivers, lakes, and other surface waters); cadastral (land ownership) data; and natural resources data (geology, ecosystem distribution, soils, and wetlands). The Geospatial One-Stop initiative is developing standards for seven of the aforementioned data types (excluding natural resources), confirming their continued importance to government. Given government buying power, it is not

---

[1]As with any taxonomy, there are ambiguities. The physical and human worlds are intertwined. Human actions influence physical processes and patterns.

[2]NRC, 1995, *A Data Foundation for the National Spatial Data Infrastructure,* Washington, D.C., National Academies Press.

[3]*Geodetic control* refers to the common reference system for establishing coordinate positions (e.g., latitude, longitude, elevation) for geographic data.

surprising that geographic data markets tend to produce the data types in which government is most interested.

3.3 STRUCTURE OF THE GEOGRAPHIC DATA MARKETPLACE

The geographic data marketplace is a network of government suppliers, commercial suppliers, value-added intermediaries, and users (Figure 3-1). This range of players is much broader than it was 30 years ago, when data capture and production were largely monopolized by government mapmakers.

Technological advances, coupled with increasing demand for geographic information and analysis, have reduced barriers to entry into the market. Consequently, the commercial sector is vibrant and competitive. Commercial firms generate a wide variety of products and services, including off-the-shelf data and information products, for-hire data acquisition services, and custom data processing services. In addition to data, firms may offer specialized access and analysis tools, and/or Web-based services that support a variety of applications including in-vehicle navigation, location-based services, Web mapping, and asset tracking. This diversity of data products and services continues to grow, to the benefit of users.

At the most basic level, government (the public spatial data infrastructure[4]) and commercial organizations (commercial data suppliers) (Figure 3-1) provide data collection and basic geographic information layers. Commercial suppliers also supply more specialized layers for private clients. On the government side, NASA, NOAA, NRO, and USGS[5] operate Earth-observing satellites. NOAA, NASA, and other government agencies also operate aircraft that collect imagery. Various public agencies also create both basic and mission-specific geographic information. For example, USGS creates hydrology layers using in-house staff and private contractors. Most federal agencies make such information available to the public at the cost of reproduction as paper maps or digital data layers.

[4]A public spatial data infrastructure comprises more than solely public data suppliers. It also includes a data infrastructure investment by many public agencies in standards, clearinghouses/metadata catalogues, online content, training, public spatial data policies, coordination across agencies, and activities such as *The National Map*, Geospatial One-Stop, and the 133 Urban Areas project.

[5]NASA is the National Aeronautics and Space Administration, NOAA is the National Oceanic and Atmospheric Administration, NRO is the National Reconnaissance Office, and USGS is the U.S. Geological Survey.

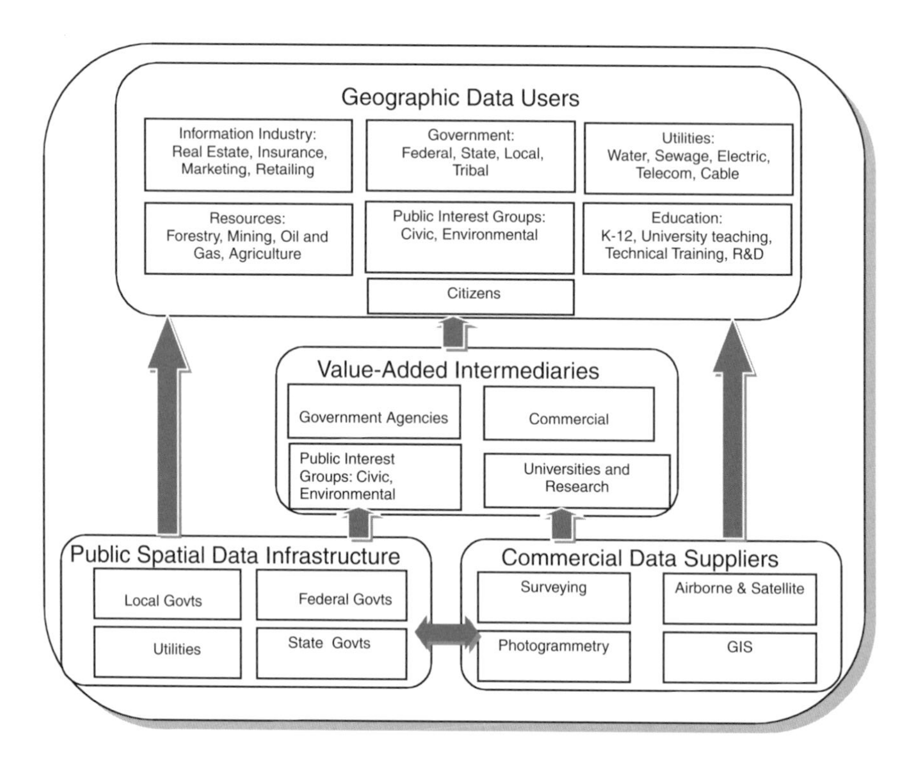

FIGURE 3-1 Geographic data marketplace. The concepts of the *geographic data market* and the *public spatial data infrastructure* are distinct. Government resources provide an infrastructure from which the marketplace for geographic data products and services emerges. Government also plays an important role inside the marketplace through its procurement of geographic data products and services. SOURCE: Adapted from: X. R. Lopez, 1996, Stimulating GIS innovation through the dissemination of geographic information, *Journal of the Urban and Regional Information Systems Association* 8(3), 24–36.

On the private side, several commercial companies operate Earth-observing satellites or aircraft. Additionally, numerous companies create basic and client-specific information layers under contract or as licensed products.

Value-added intermediaries provide additional value to users by enhancing preexisting public and private information. These enhancements include creating market-specific information layers, developing information access and analysis tools, and providing Web-based services. Intermediaries also combine datasets to meet user requirements, and improve the detail, accuracy, and precision of underlying datasets.

The commercial sector is a major provider of value-added products and services. For example, Navteq, Inc., adds value to publicly created geographic information sets (i.e., by including additional information, correcting errors, and keeping the information current) and bundles them with information-access software for in-car navigation. Researchers and government agencies also engage in value-adding activities. For example, the USGS bundles Landsat imagery, hydrology, digital line graphs, topography, land use, and other basic layers with commercial Web-access software (ArcIMS [Internet Map Server]) to create elements of *The National Map*.[6] Public interest groups may also add value by analyzing and repackaging data to highlight a trend or issue.

Intermediaries also help government disseminate information. Traditionally, the principal intermediaries were nonprofit libraries and government depositories. Today, commercial data vendors, scientists, academics, and other government agencies use the Web and online search capabilities to provide government data to users. Alternatively, some agencies exploit the Web to eliminate intermediaries altogether.

Geographic data uses tend to be demand driven. Users increasingly want data or combined data/software offerings tailored to their own specific problems.

3.4 THE GEOGRAPHIC VALUE CHAIN

The geographic offerings and affiliated services described in the preceding section flow through numerous public and private organizations to form the geographic value chain. The flow is governed by various agreements, including licenses, and is enhanced by standards.[7] This section describes the evolution of the value chain and the levels within it.

---

[6]*The National Map* "provides public access to high-quality, geospatial data and information from multiple partners to help support decisionmaking by resource managers and the public." See <http://nationalmap.usgs.gov>.

[7]For example, the Open GIS Consortium has developed specifications that have enabled growth in the Web services marketplace for geospatial applications, and the Federal Geographic Data Committee has coordinated the development of more than 30 standards for frequently used government data.

3.4.1 Evolution of the Value Chain

Until the 1980s, maps existed only on parchment, vellum, or paper. Reproducing a map was difficult, and analysis of geographic interconnections between different maps was practically impossible. During the early 1980s, however, mapping technology evolved with the broader computer revolution, and digital maps emerged. The difficult task of making physical copies became easy, facilitating sophisticated data manipulation and geographic analysis across space and time.

These changes affected the way that people work. For example, firefighters traditionally relied on paper maps of vegetation, roads, and topography. They used their knowledge of fuel types, weather, and terrain conditions to estimate how a fire would spread. Today, they call on digital information and supporting computer projections to create detailed forecasts of how fire will burn across space and time.

The range of data and services available to society has grown rapidly over the past two decades. Today's value chain embraces the full range of activities, from raw data collection to mapmaking to query and analysis tool development and Web-based services.

3.4.2 Components of the Value Chain

We identify seven levels within the value chain for geographic data and services (Figure 3-2), starting with initial data collection and processing, and moving upward to Web-based services.[8]

1. *Data Collection and Processing*. This level includes gathering and initial manipulation of raw imagery (e.g., optical, radar [radio detection and ranging], and lidar (light detection and ranging), Global Positioning System (GPS) points, or other types of data (e.g., demographic, economic, or health data). Because many of

[8]Not shown in Figure 3-2 or discussed in this section is product differentiation by scale or resolution. This adds a third dimension to the value chain. For example, Landsat and Orbital Image Corp. have very different products. Although each offering occupies the same level in the value chain, the former has 30-meter spatial resolution compared to 1 meter for the latter. Needless to say, their prices and user groups are very different. For a discussion of pricing differentiation techniques for information products and services, consult C. Shapiro and H. Varian, 1998, *Information Rules: A Strategic Guide to the Network Economy*, Boston, MA, Harvard Business School Press.

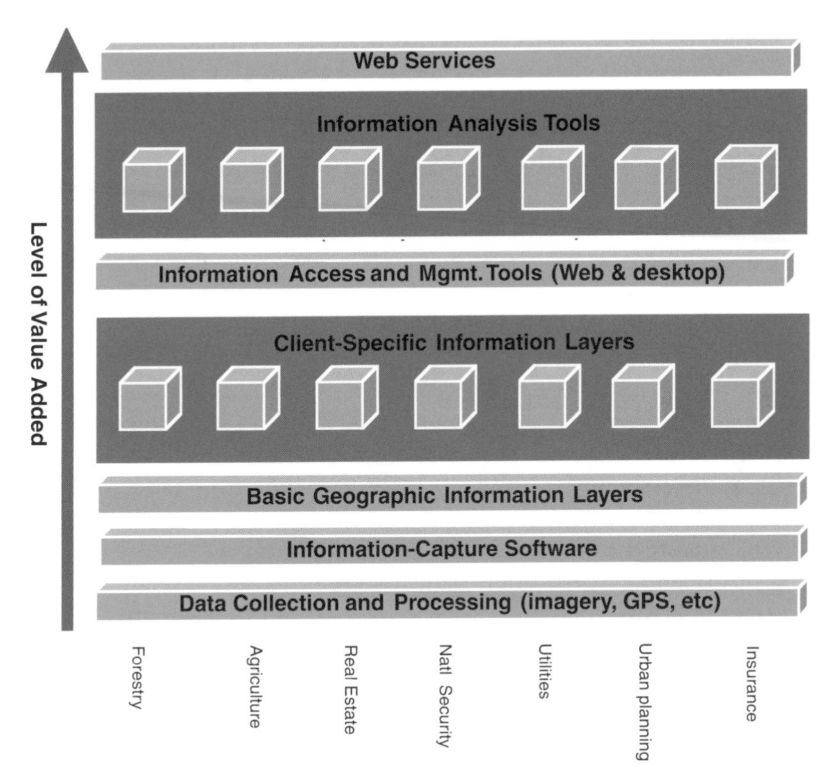

FIGURE 3-2 The geographic data value chain. The vertical axis represents the progression (from bottom to top) of the increasingly complex offerings described in text. The horizontal axis carries examples of typical markets for geographic data products and services. Boxes that span the figure illustrate levels in the value chain where market demand is sufficiently high to support standard offerings across multiple markets. Other levels of the chain, illustrated by isolated square boxes, focus on particular individual markets.

the foundation and framework data layers used by government (e.g., topography, transportation, hydrology, land use, and land cover) are derived from satellite and airborne imagery and GPS data, this report focuses on these data. Imagery collection tends to have high fixed costs, including up-front investments in aircraft, sensors, or satellites. On the other hand, imagery provides a base for information collection across a variety of markets. This means that firms can usually offer imagery capture services and products across multiple markets.

2. *Information-Capture Software*. This software lets users create digital geographic information layers by interpreting imagery or GPS data, inputting data sources (such as demographic datasets), or scanning or digitizing paper maps. Imagery and raw data are converted into geographic information system (GIS) layers through computer-aided manual interpretation or automated image classification.

3. *Basic Geographic Information Layers*. This level includes many of the information types most useful to government. These foundation and framework layers (e.g., transportation, land use, topography) are often generated using information-capture software to interpret raw data. Like data collection services, basic geographic information layers often span multiple markets (Figure 3-2). For example, transportation maps showing addresses and street names may be useful to private motorists, disaster response agencies, delivery companies, insurance agencies, and school districts. Theoretically, the existence of these multiple user groups can support speculative investment in the base layer.

4. *Client-Specific Information Layers*. Many clients have unique needs. In these cases, the market for data is limited. For example, detailed timber type maps of a particular lumber company's classification scheme offer little value to other groups. Such maps are more likely to be produced through a one-off, for-hire data acquisition and processing service. In this situation, licensing—with its emphasis on multiple licensed users—is irrelevant.

5. *Information Access and Management Tools*. To be useful, geographic data must be accessible through desktop-computer, Web-based, or mobile applications. Several private and public organizations have created powerful offerings by bundling information-access or -processing software with public information to meet needs that span multiple markets (see Box 3-1).[9] Access tools

---

[9]For example, most of the commercially successful digital map products sold to the public.

such as Space Imaging's GeoBook,[10] also serve multiple markets because the underlying information management and query algorithms usually can be ported from industry to industry at minimal cost.

BOX 3-1
Examples of Information Access Tools[a]

Commercial firms have developed a wide variety of innovative information access tools:

- DeLorme's professional mapping software and data suite provides instant access on a desktop computer to USGS digital orthophotographs, topographic maps, digital elevation models, and high-resolution satellite imagery.[b] Although these information products are also available from USGS, DeLorme's software increases their value by making them easier to search and manipulate.
- AirPhotoUSA provides online viewing of imagery that covers the United States at levels of spatial resolution ranging from 30-meter Landsat Thematic Mapper data of the entire country to 1-meter airborne data of cities.[c] Visitors to the site can download information, order hardcopy prints or CD-ROMs, or view data online.
- TerraServer is a popular Web site that couples information access software with data.[d] This site facilitates searching and purchasing imagery of various scales and sources worldwide.
- The Environmental Systems Research Institute's Geography Network hosts a popular online facility for finding, downloading, or ordering geographic data from wide-ranging sources.[e]
- Innovation is not confined to commercial firms. USGS is creating information access tools as it develops *The National Map*.[f]

[a]The examples listed are drawn from comments submitted to the committee and are meant to be illustrative rather than representative of the full range of information access tools available.
[b]Ten-meter spatial resolution SPOT (Systeme Probatoire Pour l'Observation de la Terre) imagery.
[c]Available at <http://www.airphotousa.com>.
[d]Available at <http://www.terraserver.com>.
[e]Available at <http://www.geographynetwork.com>.
[f]Available at <http://nationalmap.usgs.gov>.

[10]Available at <http://www.spaceimaging.com/solutions/geobook/>.

6. *Information Analysis Tools*. Perhaps the most sophisticated uses of geographic information are to predict future events (e.g., how wildfires will burn) and prescribe future actions (e.g., how a community should minimize wildfire risk, given limited resources). Information analysis tools for these tasks tend to be highly specific, and are usually aimed at a single user or group of users. Such tools usually are built on top of standard geographic database management software (such as those licensed by IBM, Intergraph, Microsoft, Oracle, and Sybase). Firms typically sell management software off the shelf or offer consulting services to users who want to build their own analysis tools.

7. *Web Services*. Web services[11] for mapping and geospatial analysis are a recent development and have adopted several models. One widely used Web-based mapping service is MapQuest.[12] Visitors to MapQuest's site can obtain directions and maps to any address in the United States, Canada, and several other countries. The site supports itself by posting advertisements from restaurants, hotels, and other attractions. This revenue model is relatively simple. Since MapQuest is a "stovepipe" application, it does not link with data from other Web-based mapping applications and faces few licensing or copyright issues.

FEMA's Hazardmaps Web site[13] and the New South Wales Resources Atlas[14] represent a different approach. These Web services employ open standards to facilitate rapid access and integration of geographic data from various sources on the Web. The open standards also make it easier to access and apply services provided by other developers or vendors across the Web.[15]

---

[11] *Web services* are self-contained, self-describing, modular applications that can be published, located, and invoked across the Web. Web services perform functions that can be anything from simple requests to complicated business processes. Once a Web service is deployed, other applications (and other Web services) can discover and invoke the deployed service (Open GIS Consortium [OGC] On-Line Glossary, at <http//:www.opengis.org>).

[12] Available at <http://www.mapquest.com>.

[13] Available at <http://www.hazardmaps.gov>.

[14] Available at <http://atlas.canri.nsw.gov.au>.

[15] Any data server based on open standards can be accessed by any service using the same open interface. The decision to share data is strictly a decision on the part of the owner of the data. Typically, a Web service is "published" to a

Developments in geospatial Web services reflect rapid growth in information technologies in this area, and a new breed of geospatially enabled Web services is emerging that lets users combine geographic data and analysis tools from multiple providers. These technologies present complex licensing issues because users can now access multiple information repositories and/or service providers. Each repository may have its own licensing and copyright rules, and multiple providers may demand payment.[16] To attack the problem, members of the OGC are developing an open standard for Web pricing and ordering.[17] In this approach, a "rule base" captures each provider's rules and revenue model required for a Web-services-enabled transaction. A broader approach would involve metadata for licenses appearing within or alongside standard geographic information metadata (e.g., ISO 19115 Geographic Information—Metadata[18]) to support digital rights management approaches.[19]

3.5 DATA ACQUISITION-FOR-HIRE SERVICES VERSUS DATA LICENSING

Value-chain businesses in the geographic data and services community usually follow one of two dominant business models: Under one model, users *purchase* data, with unlimited rights of use, with the provider possibly retaining the right to engage in subsequent transactions with other users. In the other dominant model, users *license* data, which restricts the uses that they can make of the data and/or their ability to transfer the data to others.[20]

Except for commercial off-the-shelf software and search/manipulation tools, the geographic information market traditionally has been based on data acquisition-for-hire services. In this model, governments and businesses

---

registry that can be searched just as catalogs search for metadata about geographic data.

[16]See the related discussion of the "complements problem" in Chapter 6, Section 6.2.1.

[17]See <http://www.opengis.org/docs/02-039r1.pdf >.

[18]Also available as OpenGIS Abstract Specification Topic 11 at <http://www.opengis.org/docs/01-111.pdf>.

[19]See Chapter 9, Section 9.3, with attention to Boxes 9-1 and 9-2.

[20]See Chapter 1, Section 1.4, for the definitions of purchase and license. Of course, the difference between purchasing and licensing can be a matter of degree, so that a license with relatively few restrictions is very similar to an outright purchase.

buy services from vendors. These services include collecting imagery or other geographic data and converting them into meaningful information. Ownership of the information typically resides in the customer.[21]

In the early and mid 1980s, commercial satellite imagery companies SPOT and EOSAT[22] began using licenses to distribute their imagery.[23] Since then, new commercial satellite companies have invariably followed the product-for-license model.

Although most airborne imagery collection companies still operate on a for-hire basis, the emergence of mass markets, coupled with falling computer hardware prices, has encouraged firms to create more and more licensed products. Many of these products are aimed at large numbers of small consumers, and bundle information access software with large geographic datasets. Examples include digital orthoimage compilations, transportation maps, and digital line graphs (i.e., line map information in digital form) (see Box 3-1).

Shifting from a data acquisition-for-hire model centered on a few large government purchasers to a mass-market, product-for-license model requires a different business organization (Table 3-1). In the data acquisition-for-hire model,[24] the service provider faces comparatively less venture risk because fixed costs are relatively low and the majority of costs are in variable labor hours. The product-for-license model[25] faces comparatively greater venture risk because the producer speculatively invests in product creation without assurance of future revenues.[26] For example, a city

[21]Ownership in this instance means that, with reference to any data to which copyright applies, the customer is the copyright holder and, with reference to any data to which copyright does not apply, the customer has exclusive or nonexclusive rights to use and make the information available to others without restriction. See also Chapter 1, Section 1.4, for the definition of ownership as applied to the licensor–licensee relationship.

[22]Earth Observation Satellite Company. See further discussion in Chapter 4, Section 4.3, under the topic of Landsat. See also NRC, 1995, *Earth Observations from Space: History, Promise, and Reality*, Washington, D.C., National Academy Press, pp 109–115; NRC, 1997, *Bits of Power: Issues in Global Access to Scientific Data*, Washington, D.C., National Academy Press, pp 121–123.

[23]Personal communication from Neal Carney, Spot Image Corp., January 2004.

[24]The acquisition of collected-to-specification aerial photography by government from the commercial sector typically has followed this model.

[25]The acquisition of commercial satellite imagery by government typically has followed this model.

[26]Although aircraft can be rented rather than purchased, the majority of costs are from items that must be purchased, such as cameras, and data and film processing systems.

government may purchase orthoimagery either by hiring an aerial data firm to collect and process the imagery or by purchasing previously collected airborne or satellite imagery under license. In the data acquisition-for-hire case, the city can specify the collection time, area, and scale; it also will own the resulting imagery. Additionally, the service provider faces very little risk on the specific transaction as long as it fulfills its contract, because the entire cost of the imagery is recovered from a single purchaser. Conversely, a company that creates consumer data products faces a significant risk that it could collect and process imagery that will generate no or insufficient future revenue to recover the investment. However, once the product is established in the marketplace, profit margins are usually higher in the product-for-license model than in the acquisition-for-hire model. The key to maintaining these margins is to restrict the ability of users to transfer the data to other users by licensing rather than selling the data.[27] Without these margins, firms would lack an incentive to risk developing new products without a proven market.

From the city government's standpoint, licensed data have benefits and drawbacks. On the benefits side, licensed products may have greatly reduced price because acquisition costs can be shared over multiple licensees. On the costs side, licensees must acquire existing product with prespecified time, area, and scale. Customers familiar with the intimacy of a service-for-hire contract also may be put off by the lower level of customer support found in the sale of a standard product under license. Furthermore, their ability to share the data usually will be restricted by the license.

In some cases, the cost advantages of licensed data should offset the disadvantages of losing downstream uses of the product and foregoing the ability to customize data specifications. In other cases, license prices are sometimes comparable to the cost of an acquisition-for-hire contract tailored to the customer's unique specifications. Licensing normally will be unattractive in these circumstances.

Distribution strategies continue to evolve as vendors experiment in the marketplace. For example, one approach is for a large portion of upfront costs to be borne by one user or a small number of users while the vendor retains the rights to distribute the data to others (e.g., Intermap's NEXTMap Britain venture). VARGIS[28] has used a business model in which the primary user pays for a majority, but not all, of the acquisition

[27]Recall that, under our definitions, purchasers are free to transfer the data to others, whereas licensees may not be.

[28]VARGIS has negotiated different licensing versus outright-purchase agreements in the District of Columbia, Virginia, Texas, and New York.

TABLE 3-1 Characteristics of Acquisition-for-Hire Service Versus Product-for-License Business Models

	Acquisition-for-Hire Service:		Product-for-License:
Characteristic	**Custom Consulting**	**Standard Data Acquisition and Processing**	**Geographic Consumer Product**
Definition	Typically one-off service engagements focused on exploring the extension of an existing technology into a new market, or a new technology into an existing market.	Data acquisition and processing services are sufficiently standardized that procedures can be codified in ISO standards and/or manuals.	Large numbers of customers license standard deliverables. Vendors rely on a high degree of automation to fill requests.
Level of standardization	Typically contractual reference to some industry standards.	Deliverables (e.g., resolution, type of imagery, currency) vary slightly for individual customers so that services cannot be completely automated.	Products are fully defined and codified. Orders are automated and human intervention is seldom required.
Amount of producer/customer interaction	High. Client and producer work together to define processes and specify results. Risky projects place a premium on trust.	Medium high. Client and producer work together to specify results.	Medium to low. Customer typically licenses product without interacting with the producer.

Reliability of offering	Low. There are no guarantees that the technology will work. The deliverable may continue to evolve as the project proceeds. Higher reliability if performance standards are required.	Medium. Procedures are standard, but deliverables vary from client to client. Furthermore, there is a risk that procedures may be misapplied.	High. Vendors realize that unreliable products can cause serious market-share loss. Clients expect and demand fitness for use. Consumer product laws ensure quality.
Required level of internal investment	None. Client pays for entire project.	Minimal. Standardized procedures are relatively inexpensive to develop.	Significant. Capital investment in software development, product definition, and management costs can be substantial.
Economies of scale	None, or acquired through sharing.	Medium. Technology and management breakthroughs frequently lead to faster and less expensive processes.	Large. Volume sales drive down per-unit costs.
Marketing	Publication of articles, speeches, and workshops.	Customer testimonials are important, as are articles, speeches, workshops, and brochures describing standard services.	Advertisements and targeted mailings.
Pricing	Estimates by hourly rate or "cost plus" contracts.	Estimates by hourly rate or fixed-fee contracts.	Catalogue pricing based on a standard price per unit.

costs in exchange for a right to distribute these data to others. The primary user usually asks for rights to freely distribute these data within its organization and to its traditional partners, and agrees not to sell the data and to try not to distribute the data further. The vendor then markets the data to others. The vendor has reduced risk in creating the datasets initially, and the primary user has more influence over the nature and extent of the data being created.

3.6 FACTORS INFLUENCING THE CONTRACTUAL TERMS OF DATA SHARING

Geographic data offerings from the commercial sector typically are differentiated by data characteristics (e.g., currency, spatial resolution, spatial accuracy, content and classification accuracy, spectral and radiometric properties, data format, ease of access and use, availability) and use restrictions. Data characteristics determine the cost of data collection and processing. More stringent data characteristics (e.g., more accurate, more current) typically require higher production expenditure, which normally is reflected in the price consumers pay. Broader use rights also tend to increase the price to the consumer, reflecting the higher value of the rights rather than any increase in cost.

Data customers lie along a continuum. Some have stringent and relatively inflexible requirements. For example, a county fire chief who needs to support emergency response services may require current, accurate, high-spatial-resolution data so that she can locate and identify structures, trees, and open spaces. She also may need to share the data with other public agencies. Unless a licensed commercial product meets these needs, she normally will rely on commercial data-acquisition services instead.

Other customers have more flexible needs. For example, a private forest appraiser may be willing to accept dated information at a low spatial resolution subject to use restrictions. Because forests change slowly, recent imagery is not especially valuable. Furthermore, the appraiser is interested in forest type rather than individual trees, and so the spatial resolution can be coarse. Finally, redistribution needs typically are limited to a single landowner.

Many consumers have missions that allow them to accept, and make tradeoffs between, a wide range of data characteristics. In such situations, the commercial sector often finds that government data have supplanted its potential market. On the one hand, flexible user requirements usually mean a large market—the type of market that might attract commercial investment. On the other, this flexibility is attractive to government because the same data can satisfy multiple governmental purposes. Commercial vendors trying to generate profits from the lower portions of the value chain (i.e., selling imagery and other geographic data gathered primarily from sensors or direct observations with little added value) often are concerned that the availability of unrestricted government data undercuts their potential markets.[29]

3.7 COMMON TYPES OF LICENSING STRATEGIES

Companies have invented many types of licensing strategies to sell their data (see Appendix D). Prominent market strategies include

- *Mass-Market Strategies.* Industry is increasingly moving toward mass-market products, and some observers believe that these products have a strong future.[30] In this environment, licensing normally requires "click-wrap" and "shrink-wrap" form agreements. Mass-market firms have repeatedly expressed interest in offering special rates to government users.[31]

- *Thin-Market Strategies.* Many firms operate in markets where would-be buyers and sellers find it hard to locate one another. Licenses that feature minimal restrictions on reuse and redistribution are often feasible in these circumstances. Aerial survey firms traditionally have followed this minimalist model. Geographic Data Technologies (GDT) extends the strategy by transferring county-scale datasets to local governments without restriction. As a practical matter, such disclosures are too fragmented to affect demand for GDT's nationwide products.[32]

- *Niche-Market Strategies.* Many companies collect large datasets and use them to create multiple, highly specialized products for transportation, navigation, automotive, enterprise/business, Internet,

[29]NRC, 2003, *Fair Weather: Effective Partnerships in Weather and Climate Services*, Washington, D.C., National Academies Press, p. 17.

[30]See, for example, testimony of Bryan Logan, EarthData Inc.

[31]Testimony of David DeLorme, DeLorme.

[32]Testimony of Don Cooke, GDT

wireless, and other specialized users.[33] Licensees typically receive broad-use and redistribution rights with these client-specific information layers. Such licenses are feasible because (1) the derived products are so specialized that it is more or less impossible to reconstruct the vendor's original dataset and (2) the derived products have few potential customers. Several firms recently have begun marketing highly specialized products to government users.[34]

- *Bundled Products.* Many geospatial product offerings bundle data and software together. In many cases, the data are publicly available and would have little value if they were sold separately. Instead, most of the value—and license restrictions—center on the software component. Some companies claim that mass-market software is the future of commercial geospatial technology.[35]

- *Transactional Services.* Some firms add value by assembling the permissions needed to make new products from existing data. Some of these companies sell the resulting products themselves,[36] whereas others work as consultants.[37] For example, the U.S. Census Bureau hired Harris Corporation in 2002 to obtain street-centerline data. Harris plans to obtain much, though not all, of the data through negotiations with local governments.[38]

---

[33]Navteq, Inc., sells to automotive, enterprise/business, Internet and wireless, and government customers (testimony of Cindy Paulauskas, Navigation Technologies Corp.); licensing has a strong future in transportation, navigation, and other niche applications (testimony of Bryan Logan).

[34]Examples include Navteq, Inc. (testimony of Cindy Paulauskas), DeLorme (testimony of David DeLorme), and AirphotoUSA.

[35]See, for example, testimony of David DeLorme.

[36]Testimony of Don Cooke describing how GDT assembled a nationwide geographic database by obtaining rights to county- and local-scale datasets.

[37]Testimony of Chris Friel describing a project in which a geology consulting firm obtained permission to use and combine GIS software, database software, e-commerce software, geology data, USGS maps, insurance casualty data, and street-centerline information to develop a new property insurance estimator product. Chris Friel also testified that his project was hampered by the fact that the data that he needed were owned by different entities and that he needed permission from all of them to produce his product. We discuss this "complements problem" in some detail in Chapter 6, Section 6.2.1.

[38]Testimony of Robert LaMacchia, U.S. Census Bureau.

- *Guaranteed Revenues.* Many firms use guaranteed commitments from one or two large customers to manage risk in the broader market. In return, large customers usually receive significant bulk discounts or generous redistribution rights. In principle, such licensing strategies can encourage investment and provide data that might not otherwise exist to both agency personnel and the private sector.[39] Guaranteed revenue models are particularly important in the satellite industry. Examples include National Geospatial-Intelligence Agency's (NGA's) Clearview agreement, and the U.S. Department of Agriculture's (USDA's) SPOT and Earthsat agreements.[40] In these cases, payments from a single large user cover the cost of production, ensuring that the data are produced, and they permit the data to be widely disseminated to users who pay little or nothing for them.

  An alternative example, from the airborne industry, is Intermap's NEXTMap survey of Britain, for which a small number of users paid a substantial portion of the upfront costs but Intermap retained the right to sell the data to other users.[41] This arrangement has the benefits of ensuring that data that are highly valued by some users are produced but that they are also available to other users, to whom the data may have less value, at lower prices.[42]

- *Unrestricted Use.* Unrestricted use and redistribution rights become feasible at the limit where guaranteed revenues cover a large proportion of the vendor's investment. In these circumstances, vendors are often willing to take a calculated risk that buyers will not go into competition with them.[43]

---

[39]As such, these guarantees sometimes are also interpreted as subsidies that could have positive impacts on public and private sectors.

[40]The Clearview contract gave industry a large guaranteed purchase in exchange for a 75 percent per-unit price cut (testimony of Gene Colabatistto, Space Imaging); USDA has obtained volume discounts from SPOT, Earthsat, EOSAT, and Space Imaging (testimony of Glenn Bethel, USDA).

[41]Testimony of Michael Bullock, Intermap Technologies, Inc.

[42]As we discuss in Chapter 6, Section 6.2, guaranteed revenue arrangements in which a single user pays all or most of the costs of production while other users obtain the data at prices that are no greater than the marginal cost of distribution are likely to do reasonably well in achieving efficiency in both the production and distribution of information.

[43]Although not unrestricted, NGA's Clearview agreement reduced vendors' need to resell data by providing a large commitment over three years (see Appendix D, Section D.3; Roberta Lenczowski, NGA, personal communication,

- *Custom Agreements.* For large transactions, most vendors are willing to write specialized contracts that tailor use and redistribution rights to individual needs. Custom agreements deliver value by ensuring that the customer does not acquire more rights that it needs.[44]

3.8 EXCHANGE RELATIONSHIPS OF THE PUBLIC-SECTOR MARKETPLACE

Agencies rely on a wide variety of transactions to acquire and distribute geographic data (Figure 3-3), including memorandums of understanding between agencies, a variety of product-based contracts and licenses, and consulting or data provisioning service contracts. Figure 3-3 does not attempt to characterize all possible arrangements, but it illustrates many of the typical relationships between players upstream and downstream of government. Data sources on the upstream side of government include at least five readily identifiable groups:

1. *Satellite data providers* have traditionally used restrictive licensing strategies to support large upfront technology and launch costs. Although firms historically have focused on selling licensed data to the defense and intelligence communities, firms are increasingly willing to negotiate broad reuse and redistribution rights on some types of data.[45]

2. *Airborne data providers* traditionally have followed a consulting services model in which customers receive all ownership rights,

---

2003). Space Imaging would probably agree to unlimited redistribution rights under appropriate circumstances (testimony of Gene Collabatistto).

[44]Vendors usually want to know the customer's business plan before setting prices (testimony of Chris Friel, GIS Solutions Inc.); vendors who understand the customer's application can usually offer a better price (testimony of Cindy Paulauskas). As we discuss in Chapter 6, Section 6.2, price discrimination, where different users pay different prices for the same information, may be necessary for efficient production of information such as geographic data. At the same time, users may be unwilling to reveal to the seller the true value they place on the information in order to reduce the price that they actually pay. See Chapter 6, Section 6.2.1, for a discussion of the underreporting problem.

[45]See, for example, a discussion of the Clearview contract negotiated by NGA with Space Imaging and DigitalGlobe (Appendix D, Section D.3).

including intellectual property rights, to acquired data. Data obtained under these service arrangements often are treated as "works for hire" that vest copyright ownership in the customer.[46] However, several airborne companies (e.g., AirphotoUSA, DeLorme, MapTech, and Navteq) have recently experimented with licensing data to government.

3. *Cartographic data providers,* including land ownership parcel conversion firms, add value to basic datasets. Most transfer all rights to customers. However, some cartographic firms license data to customers, particularly in cases where they are called on to provide data maintenance services on a recurring basis.

4. *Speculative product or online service providers* develop data or online database services in hopes of making multiple sales. Data and services typically are licensed so that the provider retains underlying ownership.

5. *Government agencies* often are required to provide public access to their data under state Open Records laws and the federal Freedom of Information Act (FOIA). However, some local government agencies, through exceptions to FOIA, sell data under license at fees that are well above the marginal cost of distribution. Agencies license data for nonmonetary reasons as well. These include providing an inducement for collaboration, protecting data security, ensuring data timeliness, and, at times, guaranteeing credit and attribution.

On the downstream side of government, there are two levels of users: secondary and tertiary (Figure 3-3). Secondary users have access to government data and use them directly. Some secondary users may make relatively few changes to the data; others add extensive value to meet their needs. Tertiary users are downstream users who receive government data through intermediaries who may or may not have made major changes to the data. Licensing controversies often center on restrictions that limit government data availability to secondary and tertiary users.

---

[46]One long-lived exception is the sale of digital elevation models (DEMs) used to create orthoimages from raw data. Because of the cost to develop DEMs, and their residual value, airborne firms often will retain intellectual property ownership of them.

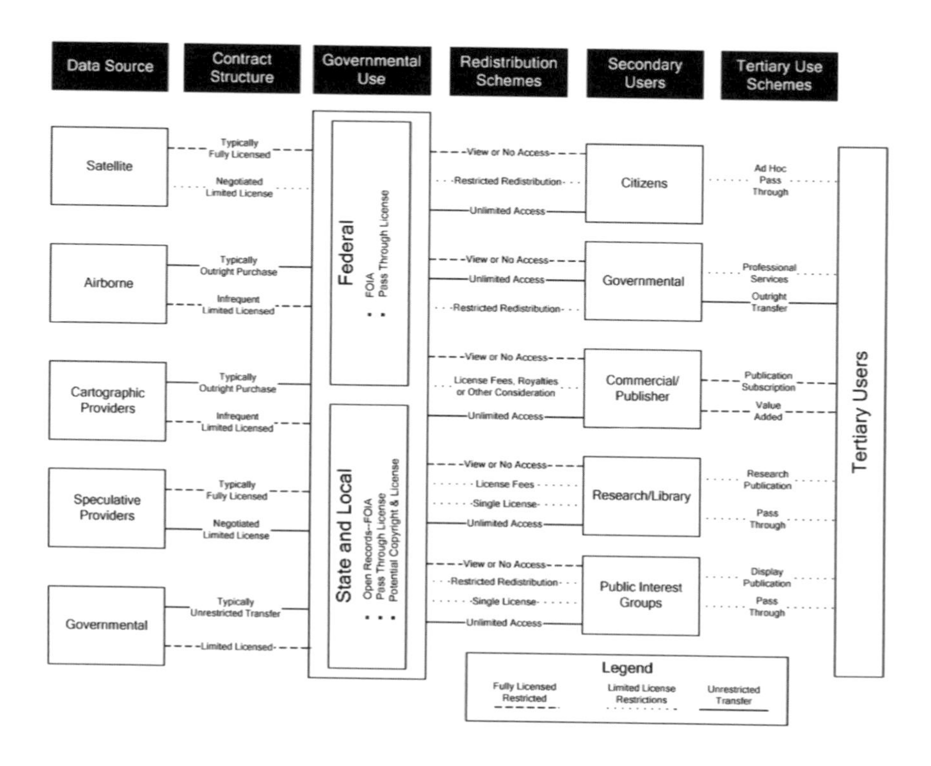

FIGURE 3-3 Conceptual diagram of geographic data contractual arrangements for data flowing to and from government.

The report recognizes five groups of secondary users, each of which can, and often does, distribute data to tertiary users:

1. *Citizens* commonly seek public data to have "access to the means of decision making"[47] or to satisfy some personal question or need. Because their motives are noncommercial, citizens generally expect to obtain public data at the marginal cost of reproduction with few or no restrictions. At the same time, citizens are usually end users: That is, they seldom redistribute data to tertiary users.

[47]E. Epstein, 1993, A case against the commercialization of public information, in Urban and Regional Information Systems Association (URISA), 1993, *Marketing Government Geographic Information, Issues and Guidelines*, Washington, D.C., URISA.

2. *Public interest groups* often advocate positions or agendas that benefit their constituents directly or indirectly. Although public interest groups generally have no profit motive, the social goals they seek for constituents (tertiary users) often have economic ramifications. Public interest groups nearly always provide data to tertiary users through some form of advocacy, publication, or display. This may change in the future as public interest groups explore making entire databases and online software processing and Web server capabilities available so that constituents can investigate data for themselves.

3. *Commercial users* (e.g., engineering firms, surveyors, developers, contract researchers, publishers) take advantage of public information in their businesses and often redistribute it. Many commercial digital mapping products (e.g., some GDT and DeLorme products) depend on public geographic data for initial product development and updates. Normally, commercial users add value by making products or services more responsive to consumer need. Other commercial firms (e.g., utilities, information firms, resources firms) use government data without redistributing them to tertiary users. Commercial users often are willing to pay fees for government data, although they obviously try to minimize costs whenever possible. Publishers often are willing to pay royalties based on sales.

4. *Nonprofit research, academic, and library* communities exist for the expressed purpose of redistributing data for knowledge advancement. Typically, they cannot afford to pay substantial royalties or fees.

5. *Government* agencies use data from other levels of government and/or other agencies to perform their functions.

Government geographic data, like all data, derive their value from use. The existence of large numbers of secondary and tertiary users makes government geographic data particularly valuable. Agencies typically distribute data as (1) the "native" or original data or (2) a "view" of the data product.[48] The distinction between these two forms (see Table 3-2) is important for licensing and agency missions. For example, FOIA (or

[48]The act of presenting a view of the data product is sometimes called "publishing," "visualizing," or providing an "image."

its state equivalent) usually requires agencies to make native data available. This provides data in the same form as that maintained by government.[49] Conversely, presenting a view of the data product seldom satisfies FOIA.

Table 3-2 Comparison of Common Modes of Government Geographic Data Availability

Characteristic	View of Data Product	Native Data
Form	Preprocessed Predetermined form and format Limited scope (possibly) Limited extent (format dependent)	Unprocessed Data in native form and format Variant-form, value-added service Full scope and extent (by request)
Audience	Ad hoc users Value-added resellers	FOIA and Open Records requests Institutional (research, public, private) Ad hoc and formal
Access mechanism	Web based Map book or map series Standard media	Standard media Web based Hard copy
Value proposition	Application services Web services Subscription services	Official public record Unvarnished content and scope
Mandate/mission satisfied	Customer service Good will Transaction cost avoidance	Meet FOIA and Open Records Statutory obligations Supports users

[49]Exceptions can occur when government puts native data in a form that is more convenient to the requester, or when government acquires data under license (in which case dissemination of these data under FOIA is subject to the terms of the agreement [Chapter 5, Section 5.4.2.1]).

Web-based viewing methods that couple data with applications shift the costs of data discovery, selection, and delivery from the agency to the user. Views also lend themselves to subscription services where users pay for value-added features.[50] By contrast, the value of native data lies in its scope, its unvarnished content, and its status as the official public record.

3.9 SUMMARY

Geographic data come in many forms and are widely used across government and society. The marketplace for geographic data is a network of government and commercial suppliers, value-added intermediaries, and users. Within this marketplace, the diversity of products and services is increasing and two dominant business models have emerged: all rights are sold to the purchaser but the vendor retains the right to use the work, or rights are retained by the vendor but customers are allowed to use the data under a license. Licensing has become increasingly common since the early 1990s, and secondary and tertiary users worry that licensing will restrict the availability of government data to them. Commercial firms worry about having to compete with government producers. The next chapter describes multiple perspectives on the role and value of licensing.

---

[50]Examples of the value-added dimension include access 24 hours a day and 7 days a week, specialized content, Web services, and reselling opportunities. The latter opportunities arise because presenting a view involves selecting and arranging information in a way that may give rise to "original expression" and thus, arguably, it is subject to copyright protection (see Chapter 5, Section 5.2.1). These opportunities only arise at the state and local levels, since federal agencies may not assert copyright ownership in works it develops (17 U.S.C. § 105).

VIGNETTE C. A SMALL BUSINESS PERSON'S DREAM

Samantha Adams runs a small business that combines information from multiple online sources to sell either as a new product or as a service to her regular customers. She is currently creating a new digital product called "My City After Dark" that includes detailed information about restaurants, entertainment establishments, and other late-night shopping. The product also includes a ghost-story walking tour of the city.

To create her product, Ms. Adams must affirmatively know that she has a legal right to incorporate the digital work products of others. Fortunately, in the new online environment, there is no need to negotiate terms of use when downloading a file. For example, Ms. Adams downloads and uses a land parcel map of the city after learning from the accompanying descriptive data ("metadata") that the local government has dedicated the data file to the public domain. She downloads portions of a restaurant directory whose metadata indicates that information may be incorporated into other products at a standard fee per item. She rejects using several comparable directories for which no metadata on use conditions is provided or for which the fees or standard use terms in the license are less favorable. She downloads several "spooky music" files that can be used for free if attribution is provided. Most of the other content Ms. Adams photographs herself or gathers from historical sources in her local public library.

The complete "My City After Dark" file is delivered back to the Web along with its metadata and licensing terms. The file and future updates may be downloaded for a stated fee by anyone, including users of personal communicators—personal mobile devices incorporating phone, e-mail, video, Web access, data processing, and location communication capabilities along with high-volume data storage. Thousands of small businesses like Samantha Adams' are now able to create similar new products or enhanced services because they can easily discover the ownership status and licensing conditions of location-based and related datasets found on the Web.

Ms. Adams' dream is one of efficient discovery, comparison, and selection of data access and use conditions. She believes that it could come closer to reality by making standardized licensing and metadata creation capabilities available to all on the Web.

4

Experiences of Government in Licensing Geographic Data from and to the Private Sector

4.1 INTRODUCTION

In this chapter, we address three of the committee's tasks: (1) exploring the experiences of federal, state, and local governments in licensing geographic data and services to and from the private sector; (2) examining the ways in which licensing of geographic data and services between government and the private sector serves agency missions and the interests of stakeholders in government data; and (3) identifying arguments in favor of and in opposition to various types of licensing arrangements.

Because the report focuses on the role of government, the chapter begins by describing federal, state, and local government experiences in licensing geographic data and services *from* the private sector. It then outlines government experiences in licensing data *to* private businesses and members of the public. The final two sections present stakeholder perspectives from private business and the academic community. Much of the material is drawn from oral and written testimony to the committee.[1] The opinions, conclusions, and arguments documented in this chapter should *not* be held to be those of the committee.

[1]See Appendix B for a list of contributors.

4.2 GOVERNMENT AGENCY EXPERIENCES IN LICENSING GEOGRAPHIC DATA AND SERVICES *FROM* THE PRIVATE SECTOR

Different government agencies have different geographic data acquisition needs. These needs should be driven by the agency's mandates, missions, goals, and operations. Mandates imposed on all agencies by state Open Records laws and the federal Freedom of Information Act and their affiliated regulations require the distribution of most government records.[2] Many agencies find it within their mission to provide data to users upon request. Inevitably, license models fit some agency needs better than others. We begin by reviewing common agency missions that help determine which types of licenses are likely to be the most useful:

- *Broad Redistribution of Data.* Agencies often collect data on behalf of society as a whole, or for the benefit of all within their jurisdictions.[3] This mission usually requires making data publicly available at marginal cost of distribution. To the extent that they exist at all, restrictions imposed when licensing geographic data from the commercial sector must be consistent with this mission.

- *Limited Redistribution of Data.* Many missions require agencies to distribute data to large groups of users. Typical numbers range from hundreds to tens of thousands of interested parties.[4]

---

[2]See Chapter 5, Section 5.4.2.

[3]For example, see the House Appropriations Committee Report 108-195 (available at <http://www.gpoaccess.gov/serialset/creports/index.html>), which states "USGS archived data are critical to Federal, State, and local governments for protecting the homeland, natural disaster assessments, and understanding global climate change."

[4]The U.S. Census Bureau must provide redistricting data to states, local governments, school districts, courts, and individuals who wish to participate in the redistricting process (testimony of Robert LaMacchia); the National Oceanic and Atmospheric Administration (NOAA) delivers data to state and local resource managers, state agencies, scientists, and emergency response personnel (testimony of Anne Hale Miglarese); the U.S. Department of Agriculture (USDA) has "hundreds" of clients (testimony of Glenn Bethel); the Federal Emergency Management Agency (FEMA) shares data with state, local, and federal agencies, and data must also be available for homeowner appeals process (testimony of Scott McAfee); Hennepin County, Minnesota, distributes data under a range of licensing arrangements to public agencies and educational institutions, and commercial companies (testimony of Randy Johnson).

- *Internal Use.* In some cases, agencies need to redistribute data to their own personnel. In cases when agencies perform their mission in cooperation with other entities, the concept of "internal use" can be stretched to include contractors, foreign governments, nongovernmental organizations (NGOs), or state, local, or tribal governments. Agencies may use licenses to limit further use or redistribution by these parties.[5]

- *Distribution of Derivative Products.* Some agencies use data to create derivative products. In many cases, there is no need to distribute the underlying data.[6]

- *Permitting Judicial Review.* All agencies must disclose data to the extent required in judicial proceedings.[7] Redistribution beyond the litigants is usually unnecessary and can be controlled by appropriate court orders.

4.2.1 Common Types of Licenses Used by Government

Agencies have experimented with a wide variety of licenses.[8] We describe prominent examples in order of increasing complexity.[9]

- *Shrink-wrap and Click-wrap Licenses.* Like their counterparts in the private sector, government employees often make use of

---

[5]The National Geospatial-Intelligence Agency (NGA) acquires some data for use and distribution only within the military and security government sectors (testimony of Karl Tammaro); the U.S. Geological Survey (USGS) occasionally procures licensed data to support "science projects and research" (written testimony of USGS, p. 3).

[6]Unlike USGS, the U.S. Census Bureau often can accomplish its mission by distributing derivative products in lieu of the underlying data with no objection to such distribution by data suppliers (testimony of Barbara Ryan).

[7]See, for example, USGS Policy 01-NMD001 (April 2001): "Even when licensed source will not be openly available, license agreements should contain terms that anticipate complications that could arise when USGS information or products are used in formulating public policy, and external parties wish to challenge USGS information products or scientific analysis through the examination of the source data."

[8]See Appendix D for a compilation of license approaches.

[9]A related discussion in Chapter 3, Section 3.7, describes broad market strategies that companies have adopted for selling licensed data.

mass-market products or online subscription services. Commercial products, such as geographic data packaged with software tools, are particularly useful when the agency mission does not require sharing or redistributing the specific data, and there is little need to acquire more specialized or detailed geographic data than that found in consumer products.

- *Embargoes and Quality Ladders.* Government agency missions do not always require the latest or most detailed data. Agencies sometimes acquire the right to redistribute data that have been embargoed for a period of weeks or years, that have been processed to impose downgraded accuracy, or both.[10]

- *Minimally Restrictive Licenses.* When use and redistribution restrictions are minimal, the distinction between licensing and outright purchase becomes academic. Government agencies that acquire unlimited rights to use and redistribute geographic data seldom demand exclusive rights. In such instances, the vendor retains the right to sell the data to others.

- *Tiered Users.* Some vendors offer "tiered" licenses in which a range of use and redistribution rights is offered. Tiered licenses let government agencies choose the option that fits their particular needs without having to negotiate a custom agreement.[11]

---

[10]The U.S. Department of Transportation licenses Geographic Data Technologies, Inc.'s (GDT's) Dynamap/2000 internally but posts Dynamap1000 (ca. 1998) for unrestricted distribution over the Internet (testimony of Don Cooke); the Census Bureau plans to license downgraded coordinates for use in its Topologically Integrated Geographic Encoding and Referencing system (TIGER) database and it will delay release of TIGER files containing licensed commercial data (written testimony from the U.S. Census Bureau); current Sea-viewing Wide Field-of-view Sensor (SeaWiFS) satellite data, the fruits of a partnership between Orbital Image Corp. (ORBIMAGE) and the National Aeronautics and Space Administration (NASA), are sold to a primary market of fishermen, whereas commercially obsolete two-week-old data are made available from NASA for use by researchers at no charge (see <http:// seawifs.gsfc.nasa.gov/ SEAWIFS/ANNOUNCEMENTS/getting_data.html>) (testimony of Anne Hale Miglarese).

[11]Levels of tiering within NGA's "traditional" license model include (in order of increasing size) single organization, Department of Defense/Intelligence Community, state/local governments, and coalition forces. The tiers (12 in total) broaden to the final tier of "Unrestricted," and then there is "Public Domain" (testimony of Karl Tammaro).

4.2.2 Nonlicense Alternatives

Agencies regularly turn to nonlicense alternatives when (1) the geographic data needed to perform their mission are unavailable; (2) alternatives are likely to cost less than procurement under license; or (3) license restrictions conflict with their missions. Most of these alternatives ultimately rely on work done by commercial firms. At the federal level, methods for acquiring geographic data or services that can accommodate but do not require licensing include use of the General Services Administration schedule, standard competitive procurement, Brooks Act procurement procedures, sole-source procurement, and memorandums of understanding with agencies that contract with the private sector.

Federal agencies also join companies in cooperative research and development agreements (CRADAs) and other partnerships for research and development, which can result in new data or technologies. Such data and technologies promote commercialization and deliver resources that neither the agency nor the private firm would likely develop on its own.[12] These alternatives may come with restrictions that let commercial partners resell the data already provided to the government agency to other users.[13]

Finally, agencies continue to experiment with a variety of innovative procurement methods. For example, USGS uses grants to obtain new data from academic researchers. Other agencies offer prizes for the development of new technologies—an approach that could, in principle, extend to database development.[14]

4.2.3 Learning by Doing

Agency license negotiation skills tend to improve over time.[15] Perhaps the simplest skill is to know how to bargain, since commercial vendors

---

[12]USGS currently has CRADAs with Microsoft (TerraServer) and National Geographic (on-demand map printing) (testimony of Barbara Ryan).

[13]NOAA negotiated a CRADA with BSB Electronic Charts to develop digital marine hydrographic maps. In retrospect, the agency has been disappointed by this arrangement, which requires NOAA and all other government and private partners to purchase Raster Nautical Charts under license from Maptech (written testimony from NOAA Coastal Services Center [CSC]).

[14]The Defense Advanced Research Projects Agency's $1 million "grand challenge" prize for advanced autonomous robots is an example. *See* <http://www.darpa.mil/grandchallenge/index.htm>.

[15]USDA's early licenses "were terrible." It took years to negotiate licenses that provide value for both sides (testimony of Glenn Bethel).

frequently are willing to negotiate on price and other license terms.[16] An agency's ability to collect data in-house or through competing vendors can provide useful leverage in these negotiations.[17] Finally, some agencies have learned to negotiate new types of licenses that potentially offer better value to the agency and commercial suppliers. For example, NGA designed its Clearview contract after carefully studying what the commercial sector needed and could accommodate.[18] Similarly, representatives of Navteq, Inc., report being impressed by government's willingness to protect the company's investment through reasonable download restrictions and indemnities.[19]

4.2.4 Coordinated Procurement

In theory, government agencies may be able to pool their purchasing power to get improved products, lower prices, or better terms when they license geographic data from the private sector. In practice, agencies must determine (1) how many agency employees will need to use the data, (2) how they plan to use the data both currently and in the foreseeable future, and (3) whether other agencies are interested in licensing the same data. Agencies have constructed ad hoc alliances to license geographic data.[20] More formal, governmentwide procurement vehicles also exist.[21]

---

[16]Space Imaging's licenses are a "first offer" (testimony of Gene Colabatistto); Maryland's Department of Natural Resources finds that private-sector companies are almost always flexible on terms (testimony of William Burgess).

[17]Vendors realize that the Census Bureau will not license data if it can collect the same information cheaper in-house (testimony of Don Cooke and written testimony from U.S. Census Bureau, p. 3); STI Services, Inc. dropped its demand for exclusive rights to Hawaiian hyperspectral data after NOAA indicated it might go elsewhere (testimony from NOAA CSC).

[18]NGA studied the private sector's needs and wrote its request for proposals in consultation with industry (Roberta Lenczowski, NGA, personal communication, 2003).

[19]Testimony of Cindy Paulauskas.

[20]For example, a NOAA/FEMA/USGS/private foundation alliance to acquire digital elevation data for southern California and a NOAA/FEMA/USGS/private company partnership to license digital elevation data for Santa Cruz and San Mateo counties (written testimony from NOAA CSC).

[21]In 1986, USGS led a governmentwide effort to license SPOT Image data. More than 30 federal agencies have made purchases totaling $42 million under the agreement (written testimony from USGS, p. 5).

4.2.5 Federal Agency Licensee Experiences

Federal agencies have experienced both benefits and drawbacks when licensing geographic data from the commercial sector.[22] Some of those experiences are presented below, followed by a summary of federal agency reactions to licensing.

4.2.5.1 Benefits to Federal Agencies of Licensing Geographic Data from the Private Sector

Federal agencies related numerous circumstances in which they received benefits from licensing. These benefits include

- *Lower Data Acquisition Cost.* Licensing frequently offers significant cost advantages over other methods.[23]

- *Immediate Availability.* Data for a specific immediate government need may already exist in the commercial sector. In other instances, private companies can quickly redirect existing acquisition systems (e.g., satellites, aircraft, or field crews) to gather needed data. When data are gathered for an immediate or urgent need, they often have a very short shelf life for government purposes, so that restrictions on reuse imposed by a commercial supplier may cause few problems for the agency.[24]

---

[22]Federal agency experiences in licensing data *to* the commercial sector are rather limited. For a discussion of factors surrounding such situations, see Section 4.3 and Chapter 8, Section 8.4.3.

[23]USDA has obtained "attractive" discounts from SPOT, Earthsat, Earth Observation Satellite Company (EOSAT), and Space Imaging (testimony of Glenn Bethel); in written testimony (pp. 1–2), USGS describes efforts to negotiate low-priced data licenses for *The National Map*; obtaining county data under royalty-free licenses "enabled the end products to be cheaper and better than they would be if FEMA had insisted on outright ownership of all the datasets" (written testimony from FEMA, p. 2). In contrast, however, private-sector firms frequently set prices that "greatly exceed the cost that would be incurred if the USGS procured the data through…some other competitive procurement process" (written testimony from USGS, p. 3).

[24]Testimony of Anne Hale Miglarese.

- *Enabling Faster Build Times.* In creating a new information system,[25] or making a new information system compatible with existing datasets,[26] it may be easier, cheaper, and quicker to use existing commercial data. Some agencies (e.g., U.S. Census Bureau) launch projects using restricted licensed data and then migrate to the use of unrestricted data over time.[27]

- *Structuring the Release of Embargoed Data.* Commercial data may lose substantial value within a few weeks or months. In such cases, licenses that restrict access or redistribution for short periods of time are often acceptable to government users.[28]

- *Supporting Specific Projects.* In many cases, government can accommodate use and redistribution limitations on data used for a particular one-time project, whereas the same restrictions would be unacceptable for ongoing operations or decision making.[29] USGS has occasionally accepted distribution restricttions on limited-coverage datasets used in specific one-time research projects.[30] Further, restrictions that would be unacceptable for distributing project results may be acceptable if applied to the underlying data.[31]

- *Enhancing Derivative Products.* In many cases, agencies use licensed data to update or correct existing government databases. License restrictions that limit government to extracting specific

[25]FEMA's Map Service Center saved "several years" by using Transamerica's restricted GeoIndex data to launch its computerized map retrieval system. FEMA plans to phase out this proprietary content over time (testimony of Scott McAfee).

[26]FEMA's experiences in Nassau County, New York, Dekalb County, Georgia, Suffolk County, New York, and Santa Cruz County, California (testimony of Scott McAfee)

[27]Testimony of Robert LaMacchia.

[28]See, for example, SeaWiFS data discussion in footnote 10.

[29]In the case of one-time projects, reduced acquisition costs are less likely to be offset by costs incurred in coordinating and administrating intellectual property rights.

[30]Testimony of Barbara Ryan.

[31]FEMA used licensed data in creating hazards assessment maps and engineer reports (testimony of Scott McAfee).

information for use in derivative products may be acceptable in these circumstances.[32]

- *Suitability for National Security Uses of Geographic Data.* Commercial data are often licensed by defense agencies in order to control public access to sensitive information.

- *A Vehicle for Allocating Risk.* In some cases, data acquirers and providers use licenses to allocate their risks if delivered products fail to perform. Examples include disclaiming warranties and limiting liability.[33]

- *A Vehicle for Proper Attribution.* Data suppliers use licenses to obtain attribution for their work. Similar terms are common in nonlicense transactions and are relatively uncontroversial.

4.2.5.2 Drawbacks of Federal Agencies Licensing Geographic Data from the Private Sector

Federal agencies report that licensing also had drawbacks compared to other procurement methods. These include

- *Product Acquisition Costs.* Licensing can sometimes result in higher acquisition costs than outright purchase. A single federal agency may have several thousand geographic data users who use the data for many different purposes. In today's market, it is sometimes less expensive and more efficient for an agency to bear the full cost of collecting and maintaining a database than to acquire and administer licenses that cover all agency users.[34]

---

[32]For example, the U.S. Census Bureau's use of CONOPS–GDT data, or its upgrade of TIGER with data from Salem County, Oregon (testimony of Robert LaMacchia).

[33]An example from a commercial firm is Intermap Technologies Inc. End-user License Agreement provided by NOAA CSC; an example from a local government is the U.S. Census Bureau's agreement provided with Marin County, California, data (testimony of Robert LaMacchia)

[34]NOAA's experience with Raster Nautical Charts (see footnote 13) is an example (written testimony from NOAA CSC).

- *Administrative Costs.* Agencies that license data from multiple vendors must keep track of the restrictions that apply to all of the data in their possession and take reasonable steps to ensure that users respect those restrictions. These challenges can be daunting for large systems with many suppliers and hundreds of users.[35] As a result, some agencies find it easier to segregate public domain and licensed data, which makes both datasets less useful.[36]

- *Transaction Costs.* Negotiating licenses takes time and effort, in part because the geographic data community is still learning how best to use this approach. Particularly for small transactions, the investment in negotiations may be prohibitive. In some—but not all—cases, the transaction costs associated with other procurement methods may require less time, money, and effort.[37]

- *Coordination Costs.* Licensing works well when agencies pool their purchasing power. However, this forces agencies to incur coordination costs to determine their own present and future needs, and the needs of other agencies that might be interested in the same data. In some—but not all—cases, the coordination costs associated with other procurement methods may require less time, money, and effort.[38] For example, if an agency acquires full rights in data there is little need to know about the detailed rights that other agencies desire since all the rights may be shared.

- *Friction Between Agencies.* Different agencies may not license the same data, or may obtain different rights to use and redistribute

[35]USDA finds that poor metadata often make it difficult to track and enforce restrictions against users (testimony of Glenn Bethel). Lack of metadata makes it difficult to know to whom Hennepin County employees can give data. Public access terminals make enforcement difficult as well (testimony of Randy Johnson).

[36]Managing licensed data in an open environment poses significant computing challenges (testimony of Peter Weiss, NOAA-National Weather Service); USDA avoids licensing data when derived products would contain both proprietary and public layers (testimony of Glenn Bethel).

[37]Robert LaMacchia testified on the U.S. Census Bureau's prolonged and ultimately unsuccessful negotiations with a commercial vendor; Scott McAfee testified on FEMA's prolonged and ultimately unsuccessful negotiations with Staten Island.

[38]Ironically, local governments often band together or form regional consortia to purchase data as an alternative to licensing data from commercial vendors (testimony of Bryan Logan, EarthData, Inc.).

the same or similar data. This incompatibility in use rights can frustrate mutual support, cooperation, and sharing among agencies.[39]

- *Loss of Public Domain "Raw Material" Effects.* Restrictions on dissemination of data obtained by government from the commercial sector can impose large costs if these data are important inputs to research and development by businesses or the academic community.[40] In such circumstances, the interests of society may be better served if agencies acquire unrestricted rights, even if this costs more than licensing.

- *Inability to Support Dissemination Needs.* Agency missions and government laws commonly require widespread dissemination of government information.[41] License restrictions that conflict with these requirements normally are unacceptable to agencies. To protect the proprietary interests of commercial companies from whom it has licensed data, a government agency may be required to enter into further licenses with all those to whom the agency has an obligation to disseminate the data. Such additional burdens may interfere with or greatly increase the cost of its dissemination obligations.

- *Ambiguous Use and Redistribution Rights.* Agencies report that ambiguous contract language sometimes deters them from using licensed data in new and unanticipated ways.[42] In theory, agency personnel can resolve such ambiguities by reading the agreement for themselves, consulting government counsel, or contacting the vendor. In practice, however, it may be difficult for agency

---

[39]*Id.* If data are owned outright by an agency, the agency is free to transfer any or all use rights to others. Presumably, friction could also occur in such a nonlicense procurement when costs are shared among agencies. However, the potential for incompatible use rights with licensed data adds an additional potential source of friction.

[40]Testimony of Peter Weiss.

[41]See Chapter 5.

[42]"It is difficult to determine what is allowed" under FEMA's GeoIndex license with Transamerica Corporation, and "innovative uses have been delayed as a result" (written testimony from FEMA, p. 8). FEMA also experienced difficulties interpreting use restrictions for Navteq data (testimony of Scott McAfee).

personnel to obtain a copy of the original license[43] or obtain advice from agency lawyers.

- *Inability to Meet Specialized Needs.* Government agencies often have specialized needs. Off-the-shelf data, which typically are offered by vendors through license arrangements, may not meet these requirements.[44]

4.2.5.3 Examples of Federal Agency Licensee Experiences

Almost all federal agencies acquiring geographic data have used licenses and will continue to do so. Examples include

- *FEMA.* The agency frequently licenses topographic data from county governments to make floodplain maps. These royalty-free licenses have allowed the agency to make maps "cheaper and better than they would be if FEMA insisted on outright ownership of all datasets."[45] Historically, FEMA has been able to accept licenses that limit redissemination to citizen appeals. Agency plans to create digital maps where users receive access to the underlying base maps and supporting data (e.g., topographic data) may put pressure on this model.[46]

- *NGA.* The agency purchases and/or licenses data for itself, the Departments of Defense and Homeland Security, and any other qualifying agency that wants to acquire geographic data through their procedures. NGA and other federal agencies also have explicit guidance to "promote stability in the U.S. commercial satellite industry."[47] Until recently, NGA acquired most images

[43]License language is seldom well documented or readily accessible in electronic form (testimony of Glenn Bethel, USDA).

[44]Testimony of Bryan Logan.

[45]Written testimony from FEMA, p. 2.

[46]Written testimony from FEMA, pp. 1–2. FEMA also licensed data from Transamerica Corp. to build a database for its Map Service Center. The data were acquired at reasonable cost and allowed the Center to open much earlier than would otherwise be possible. However, FEMA found the contract complicated to administer and ambiguous with respect to use rights. The agency is constructing a public domain database so that it can phase out the license. *Id.* at pp. 8–9.

[47]Testimony of Karl Tammaro. See also <http://crsp.usgs.gov>.

through a tiered license that offered clients 12 separate choices. In 2003, NGA introduced its Clearview and Nextview licenses,[48] which combine guaranteed purchase commitments with broad redistribution rights.

- *NOAA.* The agency believes that nonlicense procurement is the "preferred approach" to acquiring data.[49] Between 1999 and 2003, the agency negotiated licenses in four instances. In one case, the license permitted unlimited redistribution, and so, "ownership would not have added any benefit."[50] In each of the remaining cases, the agency licensed data from the commercial sector because ownership was not an option.[51]

- *U.S. Census Bureau.* The Census Bureau has agreed, and will continue to agree, to licensing agreements that do not restrict it ability to meet its mission.[52] In its extensive dealings with state, local, and tribal governments, it encounters a broad range of licenses, from those that protect against liability or address warranty issues to those that attempt to recover costs and restrict data redistribution. The Census Bureau also has tried to license data from commercial companies working with local governments.[53] In the end, it was cheaper for the Census Bureau to do the work itself.

---

[48]See Appendix D, Section D.3, for details on Clearview, and <http://www.nima.mil/ast/fm/acq/093003.pdf> for information on Nextview.

[49]"CSC believes that the general feeling within NOAA is that the licensing of spatial data acquired to meet mission requirements restricts NOAA's ability to make the best use of the information for external products and for the development of the NSDI" (written testimony from NOAA, p. 3). "Although [procurement] may appear to be more costly up front, it is the preferred approach, allowing the government to utilize the data it purchased in any way it determines to be appropriate" (*id.* at p. 8).

[50]Written testimony from NOAA, p. 4.

[51]*Id.* at pp. 4–7.

[52]Written testimony from U.S. Census Bureau, p. 3.

[53]Options discussed included purchasing "downgraded" coordinates, using only the geographic information and minimal attributes (street name only) for features not already in U.S. Census Bureau's TIGER database, and delaying the release of files with the licensed information (written testimony from U.S. Census Bureau, p. 2).

- *USDA*. The department operates a large and extensive geographic information system (GIS) program and has negotiated large-volume licenses for satellite imagery from SPOT, Earthsat, EOSAT, and Space Imaging.[54] The USDA does not license aerial data[55] but instead argues that it would be impractical to administer license restrictions for images that must be distributed to hundreds of clients.[56]

- *USGS*. The agency and its partners routinely collect and publish core geographic data without restriction.[57] However, *The National Map* program needs more data than the agency can afford to purchase. Early in 2001, USGS issued detailed guidelines specifying (1) when and under what circumstances the agency will consider licensing, (2) a detailed list of rights and exemptions that any license must grant the agency, and (3) examples of acceptable restrictions.[58] USGS has since prepared draft specifications that would allow it to procure data under license for use in making derived products.[59] USGS does not license aerial survey images on cost grounds; it contends that traditional fee-for-service acquisition methods are "far cheaper for a product that can be shared without restriction and integrated into public domain products at full resolution.[60]

4.2.5.4 Summary of Federal Agency Experiences and Reflections

When geographic data to meet a government mandate or mission do not exist, or are unavailable at a suitable price or under suitable use conditions, government agencies typically contract for new data collection.

[54]Testimony of Glenn Bethel.

[55]Written testimony from USDA, p. 2.

[56]*Id.* at p. 3.

[57]USGS's core datasets include orthoimagery, elevation, hydrography, transportation, boundary, structure, land cover, and geographic data (written testimony from USGS, p. 1).

[58]National Mapping Division Policy No. 01-NMD-01 (April 3, 2001). USGS is considering a variety of possible licensing strategies, including "generalizing the data and/or not including all of the feature attributes," embargoing data until they are "perceived to have less value to a private sector business model," and linking *The National Map* to private Web sites (written testimony from USGS, p. 2).

[59]Request for Comment 03CRR002, "Purchase of Satellite Data" (2003).

[60]Written testimony from USGS, p. 3.

Under such conditions, most federal agencies prefer acquiring full ownership rights in the data when this can be done at reasonable cost. This is based on their belief that flexibility in using such data helps support the agency's direct mandates as well as general federal mandates for access, dissemination, and duplication and waste avoidance.[61] Federal agencies typically obtain data acquisition services through commercial contracts when reasonably possible rather than gathering such data using government personnel and equipment. In achieving specific objectives, licensing sometimes can be the most effective or efficient option.

Although exceptions exist, the general framework provided by U.S. laws supports the current federal information data policy, which may be summarized as "a strong Freedom of Information Act, no government copyright, fees limited to recouping the cost of dissemination, and no restrictions on reuse."[62] Most agency personnel believe that no federal legislation or rules require them to accept restrictions on the dissemination of geographic data or accept less than full ownership rights in the data they acquire when needed to accomplish their missions. They are also under no obligation to decline such restrictions, and agencies value this flexibility to choose to license or not. Federal agency personnel generally support the current overall legal framework because they believe it gives them substantial latitude in choosing the means to acquire geographic data and services. Most agencies argue that, by whatever means, they will continue to acquire substantial amounts of geographic data from the commercial sector.[63]

Federal government reactions to licensing differ noticeably from agency to agency. However, there is a general consensus that the cost advantage offered by licenses must be weighed against constraints on current and possible future use and the interest in free exchange of information. Furthermore, the current coordination, negotiation, and administration costs associated with licensing are sometimes higher than

---

[61]See, for example, testimony from NOAA CSC, p. 3.

[62]P. N. Weiss and P. Backlund, 1997, International information policy conflict: Open and unrestricted access versus government commercialization, in *Borders in Cyberspace: Information Policy and Global Infomation Infrastructure,* B. Kahin and C. Nesson, eds., Cambridge, MA, MIT Press, pp. 300–321. For example, if a federal government agency acquires full ownership rights in a dataset and the dataset is accessible to the public as a matter of law, the government agency may not impose restrictions on the use of the data by the public, by license or otherwise, unless explicitly allowed to do so by law. See Chapter 5, Section 5.4.1.

[63]See, for example, written testimony from USDA, p. 8.

those of other procurement methods. For the most part, agencies whose missions require broad dissemination find licenses less useful than agencies that have small numbers of users or need data to make derivative products. However, some agencies assert that improved licenses and licensing practices may mitigate some of these drawbacks over time.[64]

4.2.6 State and Local Agency Licensee Experiences

Like federal agencies, state and local agencies have experienced benefits and drawbacks in licensing geographic data from the private sector. Many of those benefits and drawbacks parallel federal agency experiences and are not repeated here. Experiences of state and local governments in licensing geographic data from the private sector include[65]

- *Bakersfield, California.* Several local agencies have entered into a partnership with a private company to collect geographic data. The private partner can distribute the data under license as soon as they are collected. Government partners can use the data internally but cannot redistribute them for two years.[66]

- *Hennepin County, Minnesota.* Several local governments have signed a contract with a local engineering firm to prepare new digital orthorectified quadrangle maps. The engineering firm will prepare and maintain the maps in exchange for a fee and the right to distribute street-centerline files to consumers under license. Participating governments can use the data internally.[67]

- *Maryland.* The Department of Natural Resources (DNR) has negotiated licenses with SPOT, VARGIS, PIXXURES, and GDT. One of the licenses permits the department to redistribute the imagery, while all of the licenses allow the public to view the imagery or vector files on the agency's Internet map server,

[64]Written testimony from USGS, p. 11.

[65]See NRC, 2003, *Using Remote Sensing in State and Local Government: Information for Management and Decision Making,* Washington, D.C., National Academies Press, for additional examples of licensing experiences (in particular, pp. 22–27, 37, 47–48).

[66]Testimony of Robert Amos.

[67]Testimony of Randy Johnson.

which does not allow users to download the original image files.[68]

- *Tennessee.* Tennessee state agencies discussed partnering with a private company to make statewide parcel maps in 1998. Negotiations fell through when the company refused to include rural areas in the project and declined to provide an enterprise license for state government agencies. The agencies decided that it would be cheaper to produce the maps internally. More recently, Tennessee has licensed centerline data from GDT to build E911 systems. The license strategy has saved time and ensured data compatibility with existing datasets.[69]

- *County and Municipal Governments and GDT.* GDT has built a nationwide database by trading data with county and municipal governments across the country. These governments are usually open to novel offers, though GDT does not accept data that come with use or redistribution restrictions.[70]

Most of these experiences are not noticeably different from those discussed earlier for federal agencies. Although state and local governments are not large consumers of licensed data (e.g., in comparison to NGA), federal agencies are increasingly looking to them for data and are understandably concerned by any arrangements that limit redistribution and reuse. The balance of this section focuses on two primary approaches by which state and local bodies exchange or share geographic data in support of their missions: through coordination mechanisms and open access. These approaches often permit, but do not require, licensing or outright purchase from commercial data suppliers.

4.2.6.1 Coordination Mechanisms at the State and Local Level

Compared to civilian federal agencies, state and local government units (1) are more fragmented, (2) collect more domestic geographic data, and (3) have proportionately smaller data budgets.[71] Conditions 1 and

---

[68]Testimony of William Burgess.

[69]Testimony of Mark Tuttle.

[70]Testimony of Don Cooke.

[71]Local governments spend more on geographic data than does the federal government; two-thirds of government data are state and locally owned (testimony

3 increase the importance of interagency cooperation. Not surprisingly, state and local governments have developed various institutional frameworks to facilitate joint acquisition and sharing of geographic data:

- *Ad Hoc Collaboration.* Geographic data swapping is an increasingly common form of local-to-local transaction.[72] Additionally, local agencies may pool resources with federal, state, or other local agencies, or with private-sector partners to collect new data.[73]

- *Organized Collaboration.* In principle, state and local agencies can buy and trade geographic data among themselves through "arms-length" transactions on the open market. In practice, many governments prefer to participate in regional entities where data are shared. Examples exist in Maine, Maryland, Kansas,[74] and Illinois.[75] Some observers claim that communal organizations are more efficient than market- and license-based mechanisms for small, tightly knit user communities.[76] These organizations are less practical when disparate mandates and political structures divide potential collaborators.

- *Umbrella Organizations.* State and local governments increasingly rely on networks and partnerships to facilitate sharing.[77] For example, approximately 100 state, federal, nonprofit, and academic

---

of Scott Cameron, U.S. Department of Interior). There are currently 3,034 counties, 19,429 municipalities, 16,504 townships, 35,052 special districts, and 13,506 independent school districts (testimony of Costis Toregas, Public Technology Inc.).

[72]Testimony of Costis Toregas.

[73]See, for example, Applied Geographics, Inc., 2002, *GIS Needs Assessment and Requirements Analysis,* available at <http://gai.fgdc.gov/portal/MaineGISRequirementsReport.pdf >. This publication describes a Kansas/USDA soil mapping initiative and a Maine State Planning Office/Department of Transportation/CIO project to design a statewide GIS strategy). Additional federal/state/local/private partnerships are documented in footnote 20.

[74]See <http://gai.fgdc.gov/portal/MaineGISRequirementsReport.pdf>, p. 102.

[75]The Northeastern Illinois Planning Commission Year 2000 Digital Orthophoto Consortium brought together more than 40 local governments and state and federal agencies to acquire digital orthophotography that spanned the commission's seven-county region around Chicago.

[76]Organized collaboration is not limited to government. Utility companies routinely build shared databases that are closed to the public.

[77]Testimony of Costis Toregas.

entities in the Minneapolis/St. Paul area have created Metro GIS. Metro GIS facilitates data sharing among members, licenses data to outside users, provides a forum for exchanging best practices, and helps members coordinate their respective data collection programs.[78]

- *Contract Work.* Some agencies perform data development and processing services under contract to other government entities. This encourages agencies to pay close attention to what their users want.[79]

- *Agency Assessments.* Maine, Michigan, and Kentucky fund substantial geographic data development and project support through voluntary assessments on selected state agencies.[80] This model provides a strong incentive to pay attention to user needs in order to obtain new assessments in later years.

4.2.6.2 Sharing Geographic Data

Advocates claim that open sharing of geographic data encourages users to find and report errors, create useful reports and maps, and suggest innovative projects and improvements.[81] Detailed scrutiny and use by hundreds of citizen, agency, and business users can result in substantial benefits fed back to the agency. However, if an agency chooses to unilaterally share its geographic data openly, reluctance of other agencies to share can sometimes be a problem. Nonetheless, because practical access to an agency's geographic data can be much greater when agency personnel cooperate, requesting agencies have a strong incentive to reciprocate, and willingness to share increases over time. Arguably, some benefits of sharing data also can be realized through licensing. In some cases (e.g., Palm Beach and Broward counties, Florida), licensing

---

[78]See <http:www.metrogis.org>.

[79]Applied Geographics, Inc., 2002, *GIS Needs Assessment and Requirements Analysis*, available at <http://gai.fgdc.gov/portal/MaineGISRequirementsReport.pdf>. The North Carolina Center for Geographic Information and Analysis also develops datasets and performs analyses under contract to multiple state and federal agencies.

[80]*Id.* at pp. 92–93, 98

[81]Testimony of Robert Amos.

agreements provide incentives for formal cooperation and collaboration among parties to the agreement.

4.3 LICENSING GOVERNMENT-OWNED DATA *TO* THE PRIVATE SECTOR AND MEMBERS OF THE PUBLIC

Federal agencies have almost always distributed geographic data at or below marginal cost of distribution.[82] However, during the 1990s, many state and local governments became intrigued by the idea that their underfunded GIS programs[83] could become at least partially self-supporting. Since then, various state and local jurisdictions have experimented with licensing geographic data to others at rates that exceed the marginal cost of distribution.[84] Today, many—but not all—local government administrators believe that they have the legal authority to obtain copyrights on their geographic data, and have the right to distribute these data under license if they so choose.[85] Nonetheless, most still refrain from imposing license restrictions on reuse outside government.

Recent experiences in licensing by government to others include

- *Louisville and Jefferson County Information Consortium, Kentucky.* This local government and utility consortium maintains large-scale data on such items as land parcels, roads, curbs, fire hydrants, and building footprints. It regularly uses license restrictions to

---

[82]Congress's persistent efforts to "commercialize space" are a notable exception. We discuss the Landsat experience later in this section.

[83]Politicians usually assign a low priority to GIS (testimony of Randy Johnson).

[84]For example, state and local governments have experimented with licensing geographic data in Arizona, Maryland, Minnesota, North Carolina, and Utah (Applied Geographics, Inc., 2002, *GIS Needs Assessment and Requirements Analysis*, available at <http://gai.fgdc.gov/portal/MaineGISRequirementsReport.pdf>).

[85]Their position is controversial. Citizens recently won a lawsuit forcing Greenwich, Connecticut's city government to disclose its geographic data (see, e.g., <http://www.rcfp.org/news/2002/1030whitak.html> and subsequent articles at < http://www.greenwichtime.com/>). Additionally, Connecticut (in CT House Bill [H.B.] 5014 and CT H.B. 5039), Hawaii (in HI H.B. 443 and HI Senate Bill 427), and New York (in NY Assembly Bill 804) are all considering legislation that would restrict the public's access to geographic data (testimony of John Palatiello, Management Association for Private Photogrammetric Surveyors [MAPPS]).

- control the potential "free service bureau" role that open records treatment might encourage,
- earn a return on original consortium members' capital and maintenance investments,
- generate fee-based income for reinvestment,
- control third-party redistribution,
- require appropriate credit and attribution by makers of derivative products,
- limit liability for inappropriate use of the data, and
- limit the rights of third parties to create inappropriate derivative products.

- *Hennepin County, Minnesota.* Governments in and around the Minneapolis/St. Paul area have experimented with fee-for-service models since the 1990s. Today, their Metro GIS organization sells custom GIS services for an hourly fee. However, the operation does not cover total costs, and Metro GIS is gradually phasing out fee-based services in favor of free distribution over the Internet.[86] Nonetheless, many Metro GIS members continue to license data.[87]

- *Palm Beach County, Florida,* entered into an agreement to share portions of its proprietary geographic databases, including its most up-to-date digital orthophotography, with a consortium of insurance companies and an airborne data provider. In return, the data provider has agreed to provide the county with proprietary digital orthophotography for disaster recovery purposes within five days of a hurricane.

- *Maryland Department of Natural Resources.* Maryland DNR licensed geographic data to the private sector between 1992 and 2002. During the program's last year of operation, DNR earned approximately $7,000 in sales but spent more than $13,000 supporting sales and distribution.[88] Furthermore, sales declined during that period. Use rose 117 percent in the first seven weeks

---

[86]See <http://www.metrogis.org>. Testimony of Randy Johnson.

[87]As allowed in the Minnesota Data Practices Act § 13.03.

[88]These figures do not reflect Maryland's initial $38,000 investment in a computerized system for handling online sales (testimony of William Burgess). The DNR also distributed data on a gratis basis to partners and contractors.

after licensing ended. Staff now have more time for other activities.[89]

- *Kentucky.* Kentucky legislation allows state agencies to charge fees above the cost of distribution when the request is for "a commercial purpose."[90] Some agencies use licenses to generate these fees and limit redistribution to third parties. One example is the State of Kentucky Natural Heritage program.[91] This program distributes its biological data to NatureServe, which then adds value and resells them. A portion of NatureServe's revenue is returned to the state agency. The Natural Heritage program also provides data and analysis for specific land parcels or sites to consultants on a fee-for-service basis.

- *Landsat.* Congress has tried at least five separate funding programs for Landsat since the early 1970s. These include public, private, and hybrid schemes. From 1985 to 1992, Landsat followed a commercialization model. Despite high prices, the program failed to recover a significant fraction of Landsat's costs. At the same time, traditional users in universities, corporations, and government agencies were priced out of the market, reducing Landsat's use. Congress ultimately abandoned the comercialization program.[92]

- *Radarsat.* The Canadian Radarsat program sells synthetic aperture radar images to private users. In the past, and unlike most government fee-for-service licenses, Radarsat users received an unlimited right to produce value-added derivative products.[93] In particular, the rights were broader than those found in commercial satellite licenses, which typically prohibit redistribution of derivative

---

[89]Testimony of William Burgess.

[90]Kentucky Open Records Law § 61.870 et seq. (see L. P. Dando, 1993, A survey of open records law in relation to recovery of database costs: An end in search of a means, *in* Urban and Regional Information Systems Association (URISA), 1993, *Marketing Government Geographic Information*, Washington, D.C., URISA, pp. 5–22.)

[91]See <http://www.naturepreserves.ky.gov/heritage/>.

[92]Testimony of Joanne Gabrynowicz, University of Mississippi School of Law.

[93]This was the case in versions of Radarsat's License Agreement dated October 26, 1995, and updated October 30, 1998 (see <http://books.nap.edu/books/NI000903/html/337.html#pagetop>).

products that can be inverted to recover the vendor's original data. The current Radarsat license allows redistribution of derivative products, but with restrictions that, for example, disallow retention of the original pixel structure from their images.[94]

- *Europe.* During the early 1990s, many European governments began selling geographic and weather data. Despite high prices, revenue was negligible. High prices also seem to have stifled the growth of a U.S.-style private weather industry in Europe. European governments might have earned more through an increased tax base than they received in direct fees.[95] Over the past five years, worldwide enthusiasm for government cost recovery has declined. Many countries have either implemented or are actively considering U.S.-style marginal cost-of-distribution rules.[96]

The committee found no example of a U.S. local or state government geographic data program that covered more than a small fraction of its total GIS budget through data sales or licensing to customers outside government. In many and perhaps most cases, government sales operations fail to recover their own costs.[97]

---

[94]See <http://www.rsi.ca/about/legal/license.asp>.

[95]The value of contracts in the weather risk management industry in the five years ending in 2002 was nearly $12 billion, whereas the European market was $720 million over the same period. The difference generally is attributed to restricted dissemination of taxpayer-funded information in Europe compared to the United States. Although the European Union economy and the U.S. economy are about the same size, the United States spends twice as much on creating public sector information. The economic impact on society in terms of job creation, wealth creation, and taxes is a factor of 5 larger in the United States than in Europe (See P. Weiss, 2003, Conflicting international public-sector information policies and their effects on the public domain and the economy, *in* NRC, *The Role of Scientific and Technical Data and Information in the Public Domain*, Washington, D.C., National Academies Press).

In a parallel to the European experience, Don Cooke and Robert Amos testified that companies looking for new locations in which to invest tend to focus on areas where geographic data are freely available, at the expense of fee-for-service jurisdictions.

[96]These include Australia, China, Finland, Germany, The Netherlands, New Zealand, and Sweden. Japan already disseminates geographic data at marginal cost of distribution (testimony of Peter Weiss).

[97]B. Joffe, 2003, *10 Ways to Support Your GIS Without Selling Data*, available at <http://www.opendataconsortium.org/documents/ 10Ways_SupportGIS_Article.pdf>.

4.3.1 Government Licensing to Others for Nonrevenue Reasons

Governments also distribute data under license for reasons that have nothing to do with revenue generation. In addition to enabling proper attribution and minimizing liability,[98] these reasons include

- *Enhancing Data Security*. Licensing can provide a legal mechanism that discourages users from altering published data. This is particularly useful for protecting the integrity of data used in regulatory proceedings.

- *Promoting Collaboration*. License regimes have been used as a vehicle for organizing and formalizing collaboration.[99]

4.4 COMMERCIAL EXPERIENCES IN LICENSING GEOGRAPHIC DATA AND SERVICES TO GOVERNMENT

The views of commercial geographic data providers vary widely with respect to whether government should acquire data by license. Some commercial providers believe that licensing restrictions on government data would burden their own organizations and their clients.[100] Other companies have built their businesses around licensed data.[101] In general, companies whose business models depend on adding value to data they gather from local, state, and federal agencies are less enthusiastic about a shift toward licensing to government agencies. Data providers whose primary business models involve selling imagery or low-value-added geographic products to government tend to be more enthusiastic about licensing to government. Companies that offer packaged solutions beyond basic geographic data have not seriously pursued domestic government

---

[98]See earlier section on benefits of government licensing from the commercial sector.

[99]For example, they have been used to define "interlocal" agreements and to clarify attribution and tertiary use rules between Florida's Broward County property appraiser's office and its partners.

[100]For example, EarthData (testimony of Bryan Logan) and GDT (testimony of Don Cooke).

[101]For example, GIS Solutions (testimony of Chris Friel), Navteq (testimony of Cindy Paulauskas), and DeLorme (testimony of David DeLorme).

clients but would be more interested in doing so if government agencies were willing to accept redistribution restrictions.[102]

4.4.1 Commercial Perspectives on Strengths and Weaknesses of Licensing

Commercial licenses are still evolving. Most companies involved in licensing geographic data to government that submitted comments to the committee believe that licensing has delivered significant benefits to government. These include

- *Cost Savings.* Costs are spread among multiple customers, making data more affordable.[103]

- *Supporting Commercial Markets.* Government acquisitions encourage greater commercial investment in technology[104] and speculative data collection. This leads to a broader range of products for both government and private consumers.

At the same time, most companies recognize that licensing is less efficient and less straightforward than it could be. Problems include

- *Fragmented Markets.* Would-be users can find the satellite and USGS data they need by searching a small number of well-organized Web sites. By contrast, local geographic data tend to be dispersed among thousands of local jurisdictions. Assembling multiple datasets at the county level usually requires substantial effort. Among other effects, this condition keeps licensing from delivering benefits that would otherwise be available.

- *Excessive License Restrictions.* Negotiations often break down over complex use and redistribution restrictions. At the same time, many existing restrictions go unenforced.[105] Some companies

[102]Navteq (testimony of Cindy Paulauskas), DeLorme (testimony of David DeLorme).

[103]Testimony of David DeLorme.

[104]*Id.*

[105]Digital Globe is not aware of cheating (testimony of Shawn Thompson); EarthData is not aware of cheating (testimony of Bryan Logan); enforcement is too complex to bother with, even though many people don't delete files after

believe that simpler or more generous use and redistribution rights would increase total revenues.[106]

- *Incompatible License Rights.* Different vendors rely on different business models. Even when individual license terms are reasonable, the combined restrictions may be unacceptable. Some observers believe that greater standardization may ameliorate this problem.[107]

- *Uncertain License Rights.* Uncertain rights make licensed data less valuable to consumers. Some observers believe that clearer or more standardized rights would increase sales.[108]

- *Product Uncertainty.* Agencies seldom tell industry what types of geographic data products they plan to develop. The resulting uncertainty inhibits investment. In some cases, neither government nor industry steps in, and needs go unmet. Among other effects, this blocks the creation of commercial products that would otherwise be licensed to government.

4.4.2 Summary of Commercial Experiences and Reflections

Although some companies earn significant revenues from licensing geographic data, most revenue from sales to the domestic government agency sector and other domestic clients still comes from selling data acquisition and processing services.[109] The data acquisition-for-hire model

---

their annual subscription period ends (testimony of Chris Friel, GIS Solutions Inc.); DeLorme spends nothing on enforcement—it is more cost-effective to invest in new products instead (testimony of David DeLorme); Navteq spends a small fraction on enforcement, although it does monitor competing products for traps and audits (testimony of Cindy Paulauskas).

[106]License terms impede sales and have become “a brake on the industry” (testimony of Bryan Logan); dissatisfaction with licenses has persuaded many consumers to “make do” with public domain data or refly missions *de novo* (testimony of Chris Friel).

[107]See, for example, testimony of Chris Friel.

[108]Downstream product rights are not clearly defined, although the situation is improving (testimony of Bryan Logan).

[109]This is the sense of the committee and industry observers (personal communications, December 2003, from James Plasker, American Society of Photogrammetry and Remote Sensing; Charles Mondello, Pictometry, Inc.; and

persists because agencies and other traditional large-volume purchasers in the private sector (1) are accustomed to purchasing data acquisition services,[110] (2) perceive that they have specialized requirements that are not met by prepackaged licensed products, or (3) believe they gain the best value by acquiring full ownership rights in the acquired geographic data.

As a general proposition, agencies are not convinced that licensing offers significant overall price reductions compared to data acquisition services for large-volume airborne imagery purchases.

Commercial success with the licensing model tends to occur with data sales to large numbers of nontraditional low-volume customers who acquire imagery or other basic geographic data that were unaffordable under the acquisition service model. Examples include AirPhotoUSA (data for realtors and appraisers), and Navteq (data for transportation managers). Commercial vendors also have built successful business models based on licensing bundled packages of software tools and data.

Finally, commercial vendors recently have begun to successfully distribute, under license, value-added geographic data layers that include such information as land-use classifications and physical structures.[111] These products can sometimes offer substantial savings over traditional fee-for-service acquisition models. Licensed data can be a viable alternative when users are sufficiently flexible to accept predefined scale and classification schemes.

The geographic data industry is evolving rapidly. Some observers believe that existing product or fee-for-service models will eventually converge, making license restrictions looser but also more prevalent.[112] Other observers argue that technology eventually will drive data acquisition costs so low that it will become pointless to distribute old data under

---

John Palatiello, MAPPS), although it has not been possible to support this with revenue statements because they are not broken down in a convenient way.

[110]Navteq perceives agencies as having a "not invented here" reluctance to license outside data (testimony of Cindy Paulauskas); agencies often resist licensing, and work together to pursue alternative strategies (testimony of Bryan Logan).

[111]For example, Earthsat sells licenses to its GeoCover product with world-wide land cover. Sanborn's CitySets contains physical structures and, among many options, the number of floors and construction materials therein. The insurance industry relies heavily on these products. Insurance Services Office sells subscription services to a fire risk database in California. The service provides such information as fuel, windiness of roads, and slope. ISO also provides a subscription service to locations of hydrants and capability of fire response by jurisdiction.

[112]Testimony of Gene Colabatistto.

license.[113] Still other observers argue that technology will accelerate licensing by making existing subscription services simpler so that data can be sold as products.

For now, the only safe prediction is that licensing models will continue to evolve. Company representatives identified four trends that are likely to make licensing stronger over time.

1. *Improved Contract Design*. Some vendors have begun to promote clear, attractive licenses as a sales tool. Competition between vendors will almost certainly make licenses more attractive to consumers.[114]

2. *Validation*. Consumers are often skeptical about the quality of licensed data.[115] Some vendors believe that government or commercial-sector certification could increase sales.[116]

3. *Standard Licenses*. Licenses that were formerly negotiated on a one-off basis are becoming standardized.[117] This trend is likely to reduce transaction costs and legal uncertainty over time.

4. *Simplification of Negotiations*. Each successful negotiation provides a template for the next one.[118] For this reason, negotiating costs should fall over time. At the same time, changing

---

[113]Testimony of Bryan Logan.

[114]Testimony of Gene Colabatistto and Cindy Paulauskas.

[115]In Chapter 6, Section 6.2.1, we discuss the problems that are created by the fact that information is an "experience good," one for which the acquirers may be unable to attach a value until after they have used it.

[116]Testimony of John Palatiello and Bryan Logan. The government already certifies aeronautical and marine navigation charts.

[117]Navteq offers click-and-accept automotive licenses over the Internet (testimony of Cindy Paulauskas); testimony of John Palatiello describing MAPPS efforts to develop standard-form licenses for its members. Standardized contracts are only one means for reducing transactions costs. In Chapter 6, Section 6.2.1, we also briefly discuss the use of blanket licenses and the creation of centralized marketplaces, which can effect the same types of reductions.

[118]The effect of cumulative past behavior on costs is termed "learning by doing" by economists. Chris Friel testified how parties used an earlier agreement to move deadlocked negotiations forward, *but* consider testimony of Cindy Paulauskas that vendors may keep successful licenses secret in order to preserve competitive advantage.

technology and business models may delay the process by making earlier transactions irrelevant.[119]

4.5 ACADEMIC AND LIBRARY EXPERIENCES IN LICENSING GEOGRAPHIC DATA AND SERVICES, AND REFLECTIONS

Members of the global scholarly community have taken advantage of the inexpensive and efficient opportunities offered by digital networks to share data and knowledge among themselves with relatively few legal, policy, or technological encumbrances. Most researchers within the academic community and government support the notion that publicly funded data of interest to researchers should be openly available, absent compelling considerations and policies to the contrary.[120] As stated in an OECD report,[121] "[a]ccess to and sharing of data reinforces open scientific inquiry, encourages diversity of analysis and opinion, promotes new research, makes possible the testing of new or alternative hypotheses and methods of analysis, supports studies on data collection methods and measurement, facilitates the education of new researchers, enables the exploration of topics not envisioned by the initial investigators, and permits the creation of new data sets when data from multiple sources are combined." Geographic data are also used in education from elementary schools up to and including graduate-level research. Unrestricted access to government geographic data enhances these uses.

The interests of students, teachers, researchers, libraries, and university administrators in gaining access to geographic data are not necessarily aligned. Students and teachers may need legal and convenient access to data to accomplish class demonstrations, laboratory exercises, and class projects, but care little about the right to publish datasets or derivative products. Researchers need full and open access to the work of others, including the underlying data upon which processes have been applied or hypotheses have been tested, to test the validity of published findings.

---

[119]Testimony of Chris Friel in relation to changing technology; testimony of Bryan Logan in relation to changing business models.

[120]For a discussion of the economic perspective on publicly funded research, see NRC, 1997, *Bits of Power: Issues in Global Access to Scientific Data*, Washington, D.C., National Academies Press.

[121]OECD (Organisation for Economic Co-operation and Development), 2002, Interim Report, OECD Follow Up Group on Issues of Access to Publicly Funded Research Data, available at <http://dataaccess.ucsd.edu/Final_Interim_Report_20Oct2002.doc>.

Thus, they need the legal and practical ability to access, use, and extend datasets, including the right to publish. University library administrators have an interest in balancing the needs of all information resource demands on a campus as well as supporting the continuation of library functions. For this reason, their needs seldom reflect the priorities of any single researcher or academic group. University administrators must balance an even broader range of demands, and may be tempted to impose restrictions on the free flow of the information products of their faculty and researchers in attempts to increase their own institution's income.

Scientists and legal scholars are exploring and pursuing institutional, technological, and legal approaches designed to preserve openness and promote the advancement of science and innovation. For example, open-access electronic publishing approaches are being implemented on a widespread basis.[122] Some of these new dissemination options might be applied to geographic databases if licensing restrictions begin to encroach on the ability of scientists to access scientific knowledge.

In general, scholarly producers of academic and research materials are among the strongest advocates for the free flow of publicly funded data and information of use to the scientific community, including the data and information that the academic community itself produces. The general belief is that "government should support full, open and unrestricted access to scientific data for public interest purposes—particularly statistical, scientific, geographical, environmental, and meteorological information of great public benefit. Such efforts to improve the exploitation of public-sector information contribute significantly to maximizing its commercial, scientific, research and environmental use."[123]

4.6 SUMMARY

4.6.1 Government Experiences Licensing Geographic Data and Services from and to the Private Sector

Despite recent interest in licensing, most federal agencies still prefer full ownership rights in the data that they acquire, when this option is available. Their reasons include increased flexibility in the use of such

---

[122]See Directory of Open Access Journals, <http://www.doaj.org>. See also <http://www.arl.org/scomm/open_access/framing.html#openaccess>.

[123] P. Weiss, 2002, *Borders in Cyberspace: Conflicting Public Sector Information Policies and Their Economic Impacts, Summary Report*. Available at <http://www.weather.gov/sp/Bordersreport2.pdf>.

data, support of agency and federal mandates relating to access and dissemination, avoidance of duplication and waste, and saving money. Nonetheless, all of the federal agencies that testified before the committee have acquired commercial data under license. Their reasons vary from being able to make maps faster and more cheaply to having no realistic alternative. Reactions to licensing differ from agency to agency, although there appears to be a general consensus that any cost advantage offered by licenses must be weighed against constraints on current and possible future use and the interest in free exchange of information. In some cases, the coordination, negotiation, and administration costs associated with licensing are higher than those of other procurement methods.

Federal agencies almost always distribute geographic data at or below marginal cost of distribution. Since the 1990s, however, many state and local governments have experimented with using licenses to generate revenue from their data.[124] Ten years later, many of these entities have concluded that fee programs (1) cannot recover a significant fraction of government data budgets, (2) seldom cover operating expenses, and (3) act as a drag on private-sector investments that would otherwise add to the tax base and grow the economy. However, licenses to provide data to users may be useful to enforce proper attribution, minimize liability, enhance data security, and formalize collaboration.

4.6.2 Ways in Which Licensing Between Government and the Private Sector Serves Agency Missions and the Interests of Stakeholders in Government Data

Agency missions can be broadly grouped into those requiring broad, limited, or internal data redistribution; those requiring distribution of derivative products; and those ensuring adequate citizen access and judicial review. In addition to utilizing outright purchases of data, agencies have experimented with a range of licenses to satisfy their missions, with mixed results. For the most part, agencies whose missions require broad dissemination find licensed data less useful than agencies that have small numbers of users or need licensed data as an input for making derivative products. Over time, some agencies have learned to negotiate new types

[124]Examples included Hennepin County, Minnesota; the State of Maryland; and various European weather services. See also examples cited in Open Data Consortium, 2003, *10 Ways to Support Your GIS Without Selling Data*, available at <http://www.opendataconsortium.org>.

of licenses that potentially offer better value to both the agency and commercial data suppliers.

Commercial data vendors have a mixture of attitudes to licensing. In general, those providers whose business models depend on adding value to data gathered from local, state, and federal agencies at the cost of distribution tend to oppose government data acquisition through licensing. Those providers whose primary business models involve selling imagery or low-value-added geographic products to government generally welcome the prospect of licensing data to the government.

Academic users and producers are among the strongest advocates for the free flow of government geographic data as well as the free flow of any other publicly funded data and information of use to the scientific community. Nonetheless, the interests of students, teachers, researchers, librarians, and university administrators in gaining access to geographic data are not necessarily the same. For example, students and teachers may need legal and convenient access to data to accomplish class demonstrations, laboratory exercises, and class projects, but may care little about the right to openly publish datasets or derivative products. Researchers, on the other hand, need the legal and practical ability to access, use, and extend the datasets and work products of others, including the right to publish.

Ultimately, however, although agencies are often charged with promoting the public interest, the interests of actual and potential user groups may be discounted by agencies faced with budgetary constraints and vendors' demands.

4.6.3 Arguments in Favor of and in Opposition to Licensing Arrangements

Arguments in favor of licensing data as opposed to outright purchase include reducing acquisition costs in many instances, making data immediately available, enabling faster build times for operational information systems, structuring data release after a given embargo period, supporting specific agency projects as opposed to ongoing operations or decision-making functions, updating or correcting existing government databases, supporting national security uses, allocating risk, ensuring proper attribution, and supporting commercial markets. Arguments against licensing include increased acquisition cost in some instances; increased negotiation, coordination, administration, and enforcement costs; uncertain use and redistribution conditions; limited redistribution rights; inability to meet specialized needs; and loss of public domain effects. Nonetheless, licenses continue to evolve rapidly and are likely to improve over time. Suggestions from the commercial sector for improving and increasing adoption of licenses include promoting better contract design, encouraging validation of licensed data to increase user confidence, developing standard-form licenses, and simplifying negotiations.

Having presented the current state of licensing experiences in this chapter, we now proceed over the next three chapters to distill the legal, economic, and public interest underpinnings of U.S. data policy in preparation for the succeeding two chapters that look to future approaches to licensing and options that could address the interests of all stakeholders in geographic data.

VIGNETTE D. A SCIENTIST'S DREAM

For thousands of years, humans believed that wildfires were unpredictable and unknowable. By the late twentieth century, scientists knew better: Measure how different fuels burn in the laboratory, acquire detailed positional data on fuels and physical conditions in the field, gather details on atmospheric conditions, and after that, everything is physics and computer models. Better computer models would result in better planning for wildfires, more timely emergency response, and savings in avoided damage to property and loss of lives.

However, the models are voracious. They require detailed and up-to-date positional information on dozens of variables. Some of the needed data can be gathered by dropping "smart dust"—autonomous millimeter-scale sensing and communication devices that track temperature, humidity, barometric pressure, light intensity, vibrations, and location—into the path of an active fire. Other needed data such as slope angles, soil types, moisture in the upper soil, vegetative ground-cover mass and moisture content, wind direction, and wind speed come from such sources as digital elevation maps, meteorological stations, ground penetrating radar, and satellite images. By correlating the active burn conditions with the prefire in situ conditions using all available data, fire progression models can be greatly improved. However, there is one more needed dataset that cannot be gathered after the fact.

In this instance, the only existing preburn imagery that is sufficiently current and detailed to allow adequate vegetation mass estimates for model development is in the archives of a commercial satellite company. Dr. Karen Jones is able to quickly find the data source online. To do good science, Dr. Jones also needs permission to disseminate the results in a manner that will ensure in-depth peer review by other scientists testing her conclusions. Fortunately, in the new geographic data marketplace, companies are increasingly flexible and liberal in granting affordable use rights to basic imagery such as the preburn imagery needed by Dr. Jones and her colleagues.

The dream comes down to this: Can the geographic data market continue to shift to reasonably and readily agree to the needed license under the desired use conditions?

5

Legal Analysis

5.1 INTRODUCTION

Although many of the issues discussed in this report are questions of policy, that is, how the government *should* acquire geographic data, there are a number of legal rights, on the one hand, and constraints on the other, that affect the manner in which government *can* acquire data. This chapter examines the laws that affect government licensing of geographic data and works.

The first section discusses intellectual property law as it applies to geographic data and works, and is concerned primarily with the rights of data providers and limits on those rights embodied in intellectual property doctrines. Mainly, those limits are concerned with balancing incentives for the production of intellectual works against the interest of the public in the free flow of information.

A discussion limited to intellectual property law would be income-plete, however, since data providers are increasingly turning to licensing to preserve and enhance the commercial value of their geographic data, which are increasingly at risk of wholesale copying as datasets and products are made available online. Moreover, licensing can transfer, limit, or expand the rights otherwise conferred by intellectual property law. Thus, the second section of the chapter discusses contract law and its limits.

The third section turns to legal rules that affect the way federal agencies acquire and disseminate data. This topic is too complex for complete coverage, but we discuss the major laws and regulations that constrain agency practices relating to geographic data. Here again, the public's interest in access to information is in tension with the commercial sector's need to maintain its products as proprietary, and often with the government's own tendencies to limit access to information for cost and other reasons.

The final section turns to the situation of state and local governments as data providers and consumers. That section points out the differences not only between private and governmental data providers, but also some of the differences between the rights and obligations of state and local governments on the one hand and federal agencies on the other.

5.2 PRINCIPLES OF INTELLECTUAL PROPERTY LAW APPLICABLE TO GEOGRAPHIC DATA AND WORKS

5.2.1 Copyright

5.2.1.1 Relevance of Copyright

Although the subject of this report is licensing, this chapter devotes substantial discussion to copyright. Historically, copyright has been the most important form of protection available for works incorporating, or based on, geographic data, although the protection it affords is limited, as explained below. Consistent with this view, the purveyors of geographic data and works frequently assert copyright protection and ownership,[1] and rely on copyright to protect their interests. Licensing traditionally has been used as a means to effect the benefits of copyright protection. More recently, the limited ability of copyright to protect geographic works and data, especially in digital form, has provided a large part of the impetus to distribute data under license rather than selling them.

[1]The ownership of copyrighted works or data is a complicated subject, a detailed discussion of which is beyond the scope of this chapter. In the absence of an express (usually written) contract, ownership rights in data not previously published are governed by rules generally applicable to trade secrets, whereas ownership rights in copyrighted subject matter are governed by the "work-for-hire" doctrine (1 R. M. Milgrim, *Milgrim on Trade Secrets* § 4.02[1][a] [2003]). As between the data provider and a government agency, ownership rights in data and in any copyrights should be dealt with expressly in the contract transferring rights in the data. The purchaser or licensee of data also may wish to obtain the seller's or licensor's warranty that it (rather than its employees or subcontractors) is the owner of all rights in the data that are the subject of the contract.

The use of licensing does not avoid the need to consider copyright, however. Rather, in drafting licenses or sales contracts for information, it is important to understand that copyright law supplies the default rules for the allocation of rights in the absence of express contractual provisions. Thus, the parties to an agreement transferring rights to geographic data or works often wish to make reference to copyright principles or to contract around the otherwise applicable rules. Further, copyright principles and policies, such as preserving the public domain, may limit the restrictive terms that can be imposed in a license of geographic data.[2]

Conclusion: Because transactions in geographic data and works will touch upon both copyright and contract principles, an understanding of how copyright applies to geographic data and works is important in licensing.

5.2.1.2 Copyrightability of Geographic Data and Geographic Works

The extent of protection available under copyright is governed by several basic principles. First, as the U.S. Supreme Court made clear in *Feist Publications, Inc. v. Rural Telephone Service Co.*,[3] copyright is not available to facts.[4] This principle seems to apply to geographic data, that is, information that represents some state or condition of the physical world.[5] The judicial decisions are not entirely consistent on this point, however.[6]

The corollary of *Feist* is that compilations of facts are subject to copyright, provided that the selection and arrangement exhibit at least a

[2]See discussion of preemption in Section 5.3.1.2.

[3]499 U.S. 340 (1991).

[4]*Id.* at 345–346.

[5]See *Sparaco v. Lawler, Matusky, Skelly Engineers*, 303 F.3d 460 (2d Cir. 2002) ("To the extent that the site plan sets forth the existing physical characteristics of the site, including its shape and dimensions, the grade contours, and the location of existing elements, it sets forth facts; copyright does not bar the copying of such facts"); *Kern River Gas Transmission Co. v. Coastal Corp.*, 899 F.2d 1458 (5th Cir.) (applying merger doctrine to find map embodying contour lines and lines showing proposed location of gas line uncopyrightable), *cert. denied*, 498 U.S. 952 (1990).

[6]*See Mason v. Montgomery Data, Inc.*, 967 F.2d 135 (5th Cir. 1992) (finding copyrightability in mapmaker's selection of sources of data to incorporate into map showing land ownership).

modicum of creativity.[7] As the Court stated, "the requisite level of creativity is extremely low; even a slight amount will suffice." Thus, many works embodying geographic data will have copyright protection because their selection and arrangement will meet this minimal standard of originality.[8]

The lines between copyrightable and uncopyrightable datasets are not easily drawn, however. Consider, for example, a database containing latitude and longitude coordinates determined by the Global Positioning System (GPS) that locate such features as building corners and fire hydrants. The coordinates and what they locate are facts, and thus are not protected by copyright. Further, individual attributes describing a building or hydrant would be facts. However, a particular selection of attributes describing a building or hydrant, especially when the selection and arrangement are one of a number of possibilities, might be sufficiently original to merit copyright protection.[9] Suffice it to say that categorical statements about the copyrightability of databases of geographic data are unwise; each case must be examined closely on its own merits.

Other works that incorporate factual material, such as maps and photographic images, may contain creative expression along with factual information.[10] Maps and photographic images, for example, often have

---

[7]Thus, whereas *Feist* held that the alphabetical white-page listings of a telephone directory were not subject to copyright protection, other cases have distinguished compilations embodying more expressive choices, such as the yellow pages of the phone book, as copyrightable.

[8]See *Montgomery County Ass'n of Realtors, Inc., v. Realty Photo Master Corp.,* 878 F. Supp. 804 (D. Md. 1995) (multiple listing service that contained elements of puffery and original display was copyrightable). In *Feist*, the Supreme Court found that an alphabetical listing of names and telephone numbers lacked the requisite creativity, however.

[9]*See Mason v. Montgomery Data, Inc.,* in footnote 6. In some instances, a selection or arrangement chosen to ensure that it is sufficiently original to gain copyright protection may not be useful because it includes or excludes data based on criteria that are not sufficiently functional for its intended purposes. For example, contour lines established above an arbitrary datum with an interval that increases logarithmically would not be useful for most practical purposes.

[10]An analogy might be drawn to historical books, which contain factual material along with the author's creative expression and arrangement. Although the author's creative expression is protected by copyright, the historical facts are not, and they may be used without the author's permission. See 1 M. V. Nimmer and D. Nimmer, *Nimmer on Copyright* § 2.11[A] (2003) (hereinafter "*Nimmer on Copyright*"). The contrary view that research was copyrightable based on the labor invested, sometimes called "sweat of the brow," was overturned by *Feist*.

been found to be copyrightable. Others may extract, copy, and use the factual information contained in the work as long as the creative expression is not copied. Thus, such works, like factual databases, often are said to have "thin" copyright protection.[11]

For example, aerial photography and satellite imagery are analogous in many ways to conventional photography. Conventional photographs have been found copyrightable because of the expressive or artistic choices of the photographer, such as the selection of subject matter, framing of the image, lighting, and exposure.[12] Even though photography of natural objects and features of the landscape may involve similar creative choices,[13] such choices are not as evident in the case of aerial photography and even less so in the case of satellite images, where framing and other aspects of the image may be determined largely by the technology and practical considerations rather than by creativity. Digital maps based on geographic data are similarly likely to involve minimal expression, particularly when they are generated by computer software, using standard conventions for display of various features. As with databases of geographic data, the copyrightability of aerial photographs, satellite images, and digital maps defies easy categorization and should be assessed on a case-by-case basis.

Conclusion: Although geographic data equivalent to facts will not be protected by copyright, compilations of geographic data such as databases and datasets, as well as maps and other geographic works that incorporate creative expression, may have copyright protection. Even if copyright applies, however, copyright will not protect individual facts.

Conclusion: It is often impossible to resolve definitively whether particular subject matter is protected by copyright. When contracting for the outright purchase or licensing of geographic data or works, it is important for the agreement to address (1) whether the licensor or the seller claims copyright protection, and (2) the extent to which the parties intend to transfer or license such rights.

---

[11]D. Karjala, 1995, Copyright in electronic maps, *Jurimetrics Journal,* 35: 395.

[12]An early case involved a photograph of Oscar Wilde, which was held to have copyright protection. *See Burrow-Giles Lithographic Co. v. Sarony,* 111 U.S. 53 (1884). For a more complete analysis, see 1 *Nimmer on Copyright* § 2.08[E].

[13]Few would doubt the copyrightability of the works of nature photographers such as Ansel Adams.

5.2.1.3 Copyright in Software[14]

Software for searching and manipulating geographic data may be bundled with or provided separately from a database. A familiar example is mapping software such as that used by Mapquest. Copyright protection also extends to software used to search or otherwise manipulate data and other information, and protection extends to both source and object code.[15]

Copyright in computer programs is limited, however, because courts have held that copyright does not protect the utilitarian or functional aspects of a program.[16] Thus, functional aspects of a program that can only be implemented in a limited number of ways, including aspects that are necessary to the functioning of the program or that provide efficiencies in its operation, are not protected by copyright. Additionally, courts generally do not extend copyright protection to code sequences that represent standard practices in the industry or that are necessary for external reasons, such as interoperability with software.[17]

Conclusion: Copyright in software is likely to be "thin," precluding exact or literal copying, but less clearly covering uses that alter or transform code, or that incorporate sequences of code shorter than the entire program. In general, licenses of geographic data or works should specify any conditions on the licensee's ability to copy, modify, redistribute, or make other uses of software provided in connection with the license.

---

[14]Software additionally may be subject to patent protection as discussed below. Software also may have trade secret protection when distributed to a limited number of users under contractual restrictions, or when source code cannot be obtained by decompiling the object code.

[15]1 *Nimmer on Copyright* § 2.04[C].

[16]*Computer Associates Int'l, Inc. v. Altai, Inc.,* 982 F.2d 693 (2d Cir. 1992). Another important issue, which will not be further pursued herein, is what constitutes making a "copy" for the purposes of infringement. Some courts have held that loading a document or program into the random access memory of a computer constitutes making a copy.

[17]For a contrasting view, see *Dun & Bradstreet Software Servs., Inc. v. Grace Consulting, Inc.,* 307 F.3d 197 (3d Cir. 2002). In *Dun & Bradstreet*, the Third Circuit held that the defendant could be liable for copying only 27 out of 525,000 lines of code, and rejected the defendant's argument that its copying consisted of sequences that are "standard, stock, or common to a particular topic or that necessarily follow from a common theme or setting."

5.2.2 Limits on Copyright Protection

Traditional limits on copyright protection include the right to use uncopyrightable aspects of a copyrighted work (such as facts in a factual work, or the utilitarian features of a work, as noted above) and fair use. Fair use provides a limited right to use copyrighted material for certain purposes, including research, criticism, news reporting, and education.[18] The Copyright Act[19] and judicial precedent set forth four factors to be considered in determining whether a use is fair: (1) the purpose and character of the use (e.g., commercial or noncommercial), (2) the nature of the work (i.e., factual or otherwise), (3) the amount and substantiality of the use in relation to the work as a whole, and the (4) potential effect on the market for the copyrighted work. No factor is dispositive, although a few generalizations can be made. For example, greater latitude is allowed for use of factual works than for fictional ones, and transformative uses (as distinguished from simple copying) are also favored. The effect of the use on the market for the copyrighted work is often an important if not deciding factor.

Because some geographic data and works are factual, and because some uses might be characterized as insubstantial or might be related to research or teaching activities or other favored uses, at least some uses of copyrighted geographic data would likely qualify as fair use. For example, use of a copyrighted database for research or teaching purposes, or to verify scientific claims, might qualify as fair use. Some courts have held that reverse engineering of computer software to determine how it works is fair use,[20] although this rule is not universally followed and may be subject to waiver by contract.[21] Like the issue of copyrightability, fair use analysis is fraught with uncertainty.

The misuse doctrine is another limit on copyright that may be significant for works employing geographic data. If the copyright owner is deemed to have "misused" the copyright, the copyright will be unenforceable until the effects of any misuse have been purged.[22] Misuse usually is based on the copyright owner's attempt to extend the lawful monopoly conferred by copyright to unprotected subject matter or

---

[18]For a general discussion of fair use, see 4 *Nimmer on Copyright* § 13.05 (2003).

[19]17 U.S.C § 107.

[20]See, for example, *Atari Games Corp. v. Nintendo of America Inc.*, 975 F.2d 832 (Fed. Cir. 1992).

[21]*Bowers v. Baystate Technologies, Inc.*, 320 F.3d 1317 (Fed. Cir. 2003).

[22]*Lasercomb America, Inc. v. Reynolds,* 911 F.2d 970 (4th Cir. 1990).

activities, such as anticompetitive clauses in licensing agreements.[23] At least one court, however, has rejected the claim that a contract term prohibiting reverse engineering constitutes misuse.[24] Anticompetitive conduct may also give rise to antitrust liability (see Section 5.2.3.4).

Conclusion: Fair use and the misuse doctrine represent significant limits on the copyright owner's rights. The scope of their application is sufficiently uncertain, however, that, where possible, parties should contract for anticipated uses rather than rely on fair use doctrine or other uncertain legal doctrines to sanction the licensee's activities.

5.2.3 Patent Protection and Limits

Software and the interactive processes used to access geographic data in digital form (e.g., over the Internet) are also potentially patentable, at least under current law.[25] Patent protection typically protects the series of steps or algorithm performed on the computer, rather than specific code.

There is no fair use exemption to patent infringement. An accused infringer may defend by showing that the patent is invalid or was misused, claims that would require case-by-case analysis. Case law also has recognized a narrow exemption for research; the exemption has come under recent scrutiny, however, and its scope is unclear.[26]

Conclusion: Some software used with geographic data and works may be patented, although the exact scope of available protection is an area of the law that is still developing. When patented or copyrighted software is provided in connection with geographic data or works, the terms of its use should be addressed in a license.

---

[23] 4 *Nimmer on Copyright* § 13.09.

[24] *Syncsort Inc. v. Sequential Software, Inc.*, 50 F. Supp. 2d 318 (D.N.J. 1999).

[25] *State St. Bank & Trust Co. v. Signature Fin. Group,* 149 F.3d 1368 (Fed. Cir. 1998). In one case, a federal district court ruled that the auction Web site eBay must pay $35 million for infringement of the plaintiff's e-commerce patents. *See MercExchange LLC v. eBay,* No. 2:01cv736 (E.D. Va. May 27, 2003). For a case involving software for the compression and storage of large digital images, see *LizardTech, Inc. v. Earth Res. Mapping, Inc.*, 35 Fed. Appx. 918 (2002).

[26] *Integra Lifesciences I, Ltd. v. Merck KGaA,* 331 F.3d 860 (Fed. Cir. 2003) (noting that the Patent Act does not include an "experimental use" exemption for infringement); *Madey v. Duke Univ.*, 307 F.3d 1351 (Fed. Cir. 2002), *cert. denied*, *Duke Univ. v. Madey,* 2003 U.S. LEXIS 5045.

5.2.4 Technological Controls and the DMCA

Owners of digital geographic data, like other database owners, may use technological means to protect digital information.[27] Access controls, such as the use of passwords, are common for digital content, and often are linked with the requirement that the user accept the database provider's terms (see discussion of "click-wrap" and "shrink-wrap" licenses in Section 5.3.1). Encryption also can be used to protect content from unauthorized access. Similarly, watermarking and other technologies protect against copying or deter copying by permitting identification of copied materials. Technological means also can be used to monitor the users' activities, such as accessing, opening, and reading files.[28]

The use of technological controls to protect digital content that contains at least some copyrighted material is reinforced by new legal rights created by the Digital Millennium Copyright Act of 1998 (DMCA).[29] The DMCA creates civil and criminal penalties for defeating technological measures that control access to a copyrighted work and for providing technological means to defeat access controls and copy protection measures. The DMCA also provides for civil and criminal penalties for violations of prohibitions on changes to copyright management information included in digital works. Although the DMCA contains some exceptions to its liability regime, it is widely viewed as foreclosing uses that would otherwise qualify as fair use under copyright law.

Some states also have adopted or are considering "super DMCA" legislation, which is intended to provide legal protection against the theft of telephone and cable services. This broadly drafted legislation, however, could arguably be interpreted to prohibit the use of security technologies, such as those that conceal the origin or destination of data packets transmitted over the Internet, or with encryption and decryption, which are also widely used for security purposes. Although measures are being

---

[27]The array of legal and technological tools used by content owners to protect digital content are referred to as digital rights management (DRM) (B. Frischmann and D. Moylan, 2000, *Berkeley Technology Law Journal* 15: 865).

[28]J. E. Cohen, 1996, A right to read anonymously: A closer look at "copyright management" in cyberspace, *Connecticut Law Review* 28: 981, 983–987.

[29]Pub. L. No. 105-304, 112 Stat. 2860 (1998). A summary of the provisions of the DMCA can be found at <http://www.copyright.gov/legislation/dmca.pdf>. See also 3 *Nimmer on Copyright* § 12A.03 (2003).

taken to revise the model legislation, it is not clear whether these changes will become law.[30]

Conclusion: Owners of geographic data and works continue to use technological controls to protect digital content. The DMCA reinforces these means with civil and criminal penalty provisions that override activities that might otherwise qualify as fair use. When licensing geographic data or works in digital form, agencies should include adequate provision for the anticipated downstream uses that otherwise may be precluded or called into question by the DMCA.

Conclusion: Some licenses of geographic data may require agencies to limit access to, and further use of, digital geographic data by third parties. When license agreements contemplate limited access by third parties, such as other agencies or members of the public, the agency must ensure that the conditions on access and use contemplated by the agreement are compatible with the technological capabilities of the agency.

5.2.5 Unfair Competition and Misappropriation

State unfair competition law may provide some additional protection against copying and use of databases created through significant investment of resources in limited circumstances. In some states, a "misappropriation" claim might be sustainable against someone who appropriates information whose value is time sensitive and uses it in a manner that lessens or destroys the value to the creator.

The misappropriation doctrine is based on the "sweat of the brow" or "industrious creation" theory.[31] In *INS v. AP*, the U.S. Supreme Court upheld an injunction against the INS for appropriating and publishing news items from the AP's bulletin boards, destroying their value to the AP. The Court justified the decision on the ground that the defendant's appropriation would otherwise destroy the incentive to invest in the gathering and publication of news.

---

[30]See State "super DMCA" anti-piracy bills seen undermining security protection, *Journal of Patent, Trademark & Copyright Law* 65: 588 (Apr. 18, 2003).

[31]This theory was rejected by *Feist* in the context of copyright, a development that raises additional questions about the viability of the misappropriation doctrine except on facts analogous to *International News Service [INS] v. Associated Press [AP]*, 248 U.S. 215 (1918). See 1 *Nimmer on Copyright* § 3.04[B][3][b].

Since the adoption of section 301 of the Copyright Act, claims for misappropriation of uncopyrightable facts usually have been found to be preempted by copyright law, except in cases closely analogous to the "hot news" scenario of *INS.*[32] Nevertheless, some geographic data have value that is time-limited and therefore might be appropriate for the hot-news misappropriation theory. Moreover, recent proposals for database protection embody some of the principles of the common law misappropriation doctrine (see also Section 5.2.6).

5.2.6 Database Protection Legislation

Since 1996, Congress has considered several proposals to create a new form of protection for databases. The *Feist* decision prompted concern that investment in databases would be discouraged because factual databases would not be protected by copyright. The impetus for database protection was also increased by the European Union's (EU's) adoption in 1996 of its Database Directive,[33] which created a new form of intellectual property in databases. Under the directive, the "extraction and/or re- utilization of the whole or of a substantial part" of a protected database without the owner's permission is prohibited. Protection under the directive nominally expires after 15 years, but this limitation may be meaningless in cases where data are continually updated, since protection is extended when a database is augmented through substantial new investment.[34]

Proposals for new U.S. legislation either track the EU Directive or adopt an unfair competition theory, limiting infringement to activities that impact the market for the database. The National Academies,[35] the American Association for the Advancement of Science, and other scientific organizations have questioned whether the need for database protection has been demonstrated, noting the robust database market in

[32]See, e.g., *National Basketball Association v. Motorola, Inc.,* 105 F.3d 841 (2d Cir. 1997). See generally *Restatement (Third) of Unfair Competition* § 38 Appropriation of Trade Values & Reporter's Note, American Law Institute, 1995.

[33]Directive 96/9/EC of the European Parliament and the Council of 11 March on the legal protection of databases. The directive itself is available at <http://cyber.law.harvard.edu/property00/alternatives/directive.html>.

[34]Of course, the original database would no longer be protected, but it might not be available separately from the augmented version.

[35]National Academy of Sciences, National Academy of Engineering, Institute of Medicine, Letter of October 9, 1996, to The Honorable Michael Kantor Secretary of Commerce Department of Commerce.

the United States even in the absence of this additional protection. These organizations also have identified many potential problems with the database proposals, including the absence in the EU model of any provision for fair use.[36] Additionally, research indicates that the EU directive has not provided a significant stimulus to database creation in the EU.[37]

At the time of this writing, Congress had not enacted database protection legislation, although new legislation was introduced in October 2003 and March 2004.[38] The National Academy of Sciences and the National Academy of Engineering presented testimony before Congress on H.R. 3261 and have written to Congress expressing opposition to the legislation.[39] If database legislation is ultimately adopted, licenses and purchase agreements for geographic data and works then would need to specifically address statutory database rights, in a manner analogous to provisions covering copyright.

Conclusion: Database legislation could significantly strengthen the rights of database developers and limit access to and use of geographic data. If database legislation passes, licenses of geographic data and works would need to address statutory database rights in a manner similar to contractual allocation of rights under copyright law.

5.3 CONTRACT LAW AND LICENSING

5.3.1 General Principles

Contract law[40] and licensing have begun to play an important, even paramount, role in protecting databases, including geographic databases.

---

[36]National Research Council (NRC), 1999, *A Question of Balance: Private Rights and the Public Interest in Scientific and Technical Databases,* Washington, D.C., National Academies Press.

[37]S. Maurer, P. B. Hugenholtz, and H. Onsrud, 2001, Europe's database experiment, *Science* 294: 789.

[38]H.R. 3261, Database and Collections of Information Misappropriation Act (introduced Oct. 8, 2003); H.R. 3872, Consumer Access to Information Act of 2004 (introduced Mar. 3, 2004).

[39]See <http://energycommerce.house.gov/108/Hearings/09232003hearing1086/Wulf1714.htm>.

[40]Contract law is largely a creature of state, rather than federal, law and the principles discussed herein are those developed under the law of the various states. In contrast, contracts to which the federal government is a party are governed by federal law (W. N. Keyes, 1996, *Government Contracts Under the Federal*

This development began several decades ago as digital media began to replace paper. It has been accelerated by the availability of digital media and the rise of the Internet as the preferred platform for managing large databases.

Traditionally, many forms of information have been published in print form. In that environment, copyright and the practicalities of printing prevented wholesale copying and republication. Readers, however, have been free to extract and use the factual information and ideas contained in a published work. Contractual restrictions on the use of factual information[41] in the print environment have been feasible only where the information provider maintained the information as trade secret or confidential and negotiated an agreement with each recipient of the information, typically requiring customers to maintain the confidentiality of the information and limit its use to specified purposes. Such agreements are seldom feasible for more than a few licenses. This trade secret information would not be published in the usual sense of the term; if it did become publicly known, trade secret protection would cease.

The shift to electronic media has made it possible for information providers to extend the licensing model to large numbers of customers. For example, data are often provided in electronic media such as compact disks (CD-ROMs), often packaged with software that allows searches and other manipulations of the information. Although this kind of packaging and sale of information is in many ways analogous to the sale of a print text, database providers have sometimes followed the model used for software licensing, in which the provider sells the disk, but licenses the information with restrictions on how it may be used. Typically, such licenses prohibit copying and dissemination to other potential users, and in the case of software, may contain terms restricting reverse engineering or decompiling of code.

Where a database is marketed to large numbers of potential customers (i.e., in mass markets), the license is packaged with the product or requires assent to the seller's conditions through a click of the mouse before the customer can access or install the information. The validity of

---

Acquisition Regulation § 33.22, Eagan, West Information Publishing Group. Thus, the Federal Acquisition Regulations (FARs), as interpreted in the courts, will govern contracts entered pursuant to the FARs, though these often may be interpreted with reference to the "general law of contracts" prevailing in most states. However, parties sometimes incorporate choice of law provisions that specify that the law of a particular state and federal law govern.

[41]Copyrighted information, of course, cannot be used without a license, irrespective of whether it is confidential.

these shrink-wrap or click-wrap licenses has been in question because the customer does not see the license—and therefore cannot consent to its terms—until after the transaction is completed. Such licenses also have been challenged as unlawful contracts of adhesion on the ground that there is no actual negotiation of terms.[42] Recently, however, courts have upheld these licenses, based on assertions that the customer could return the goods if the license terms were unacceptable, or that the customer is aware of the terms from previous transactions.[43]

Where databases are delivered over the Internet, each user can be required to pay the price for access and to assent to conditions imposed by the database provider, before access is permitted. These click-wrap licenses thus avoid some of the contract formation issues that arise with the sale of shrink-wrap CD-ROMs. Potential customers, however, still have little ability to negotiate license terms and often are faced with take-it-or-leave-it terms.[44]

Finally, contracts may be held invalid or unenforceable under conditions where the agreement is found to contravene some important public policy. Courts sometimes use the unconscionability doctrine to strike down terms imposed by a party with greatly superior bargaining power, especially when the terms are oppressive, unfair, or render other terms of the contract or other law ineffective.[45] However, even onerous terms in a consumer

[42]M. A. Lemley, 1995, Intellectual property and shrinkwrap licenses, *Southern California Law Review* 68: 1239.

[43]*ProCD, Inc. v. Zeidenberg,* 86 F.3d 1447 (7th Cir. 1996); *M.A. Mortenson Co., Inc. v. Timberline Software Corp.,* 998 P.2d 305 (Wash. 2000). Much of the following analysis, and indeed this report, assumes that courts will continue to uphold such licenses. Licenses that restrict the further use and distribution of published data are also questionable on the ground that they are preempted by copyright law. See discussion in Section 5.3.1.2.

[44]An additional development in this area is the "browse-wrap" license, to which the customer purportedly assents by browsing the Web site. The validity and enforceability of browse-wrap licenses is largely unsettled. An American Bar Association (ABA) committee is studying browse-wrap licenses and is expected to make recommendations on enforceability and other issues. See ABA working group participants formulating guidelines for "browsewrap" contract terms, 71 *U.S. Law Week* (BNA) 2662 (Apr. 22, 2003).

[45]The unconscionability doctrine has been developed under the Uniform Commercial Code (UCC), adopted in some form by all states. See L. Lawrence, Anderson on the Uniform Commercial Code § 2-302:9 (2003). The courts are divided on the applicability of the UCC to transactions involving information. See Anderson on the Uniform Commercial Code § 2-105:81, available at 2003 WL, ANDR-UCC § 2-105:81 (software). Recently, the American Law Institute adopted revisions to articles 2 (which governs sales of goods) and 2A (which governs

contract usually will be enforced if they are prominently called to the attention of the consumer. This doctrine seems unlikely to limit restrictive licensing terms in the context of negotiated licenses for geographic data and works, especially between providers and government agencies. Although onerous terms in online licenses offered to the general public might be vulnerable to attack, on the whole it seems unlikely that courts will significantly limit the reach of shrink-wrap and online licenses.

Conclusion: Licenses of data are a type of contract in which the data provider agrees to the licensee's right to access or use data, usually with restrictions as to the duration or scope of the use. Licenses (including shrink-wrap and click-wrap licenses) are a form of transaction increasingly favored by vendors for geographic data. Courts have upheld shrink-wrap licenses that protect uncopyrightable data compilations and are likely to uphold such licenses involving geographic data, although this area of the law is in a state of flux.[46]

5.3.1.1 UCC

The enforceability and interpretation of geographic data licenses may also be affected by article 2 of the UCC, adopted in some form in every state except Louisiana. The UCC governs the "sale of goods," but has not expressly defined "goods" as either including or excluding information. Additionally, licenses may not be deemed "sales" for the purposes of the UCC. Courts considering these issues have disagreed on whether the UCC governs transactions in software (where the vendor often licenses its use). The American Law Institute and the National Conference of Commissioners on Uniform State Laws (NCCUSL)[47] recently have approved a revision to the UCC's definition of goods that expressly excludes "information."[48] In states that adopt the revision, the UCC will not apply to transactions in

---

leases) of the UCC, which exclude "information" from the definition of "goods." Under the amendments, the UCC will not cover transactions involving the transfer of information. See ALI membership backs proposed revisions to sales, lease articles of UCC, 71 *U.S. Law Week* (BNA) 2744 (May 27, 2003).

[46]As with proposed database protection, there is little evidence that the increased incentives for database production afforded by licenses that restrict the use of published facts are warranted. See Section 5.2.6.

[47]These bodies draft uniform state laws and proposed revisions, which become effective when adopted by state legislatures.

[48]ALI membership backs proposed revisions to sales, lease articles of UCC, 71 *U.S. Law Week* (BNA) 2744 (May 27, 2003).

information, although it may apply to the sale of goods that also contain information (e.g., an automobile that contains a computer). The precise scope of coverage will be determined by the courts.

The UCC could be significant because it provides a number of rules that govern sales transactions when the parties fail to specify their intent. For example, if the parties fail to include provisions dealing with warranties, the UCC provides for implied warranties of merchantability and fitness for a particular use. There is also a well-developed body of law under the UCC dealing with the scope and meaning of such concepts as "unconscionability." Thus, the UCC might provide rules governing sales of geographic data where data and other information are deemed "goods" within the meaning of the UCC or where they are incorporated into a tangible product (such as a mapping service in an automobile).[49]

Conclusion: The UCC provides additional warranties and rights that potentially cover geographic data transactions. Proposed revisions would limit the UCC's applicability to geographic data.

5.3.1.2 Other Doctrines Affecting Contract Validity and Interpretation

Other doctrines sometimes invoked to limit the reach of onerous contract terms are pro-competitive policies (see discussion of antitrust in Section 5.3.1.4), the First Amendment, and preemption. Copyright preemption has been argued as a basis for invalidating contracts that purport to limit the use, copying, or distribution of publicly known facts.[50]

Preemption of contractual protection for published facts could be important in the context of licensing geographic data, particularly where the data are widely disseminated although ostensibly under restrictive licenses. For example, suppose that a government agency agreed to a license that allowed it to post data on its Web site, but required it to limit viewers' rights to disseminate or use the data further. A court might find such restrictions invalid because of the copyright statute's policy of permitting the free use of uncopyrighted factual material. To date, courts generally have rejected arguments that copyright preempts contract rights,[51] thus permitting contracts to confer protection on factual material

---

[49]Space limitations preclude a detailed analysis of contract terms allocating liability between vendor and purchaser or licensee.

[50]*ProCD, Inc. v. Zeidenberg,* 86 F.3d 1447 (7th Cir. 1996).

[51]The reasoning of *ProCD v. Zeidenberg* is not uniformly accepted, however. See *Wrench LLC v. Taco Bell Corp.,* 256 F.3d 446 (6th Cir.), *reh'g en banc denied*

where copyright would not. Nevertheless, the Supreme Court has yet to address this issue.

Conclusion: Courts sometimes invoke competition policy, the First Amendment, and copyright preemption to invalidate or limit contract terms. These doctrines could be invoked to limit the enforcement of license terms that restrict the free flow of public or widely disseminated information. In most circumstances involving licenses of geographic data, such a result would go beyond current law.

5.3.1.3 Uniform Computer Information Transactions Act (UCITA)

In 1999, the NCCUSL approved the UCITA, a model law for consideration by state legislatures. Designed to create a unified approach to the licensing of software and information, the draft has drawn criticism from many consumer and industry groups. As of early 2004, only 2 states—Virginia and Maryland—had adopted UCITA, although it has been considered in as many as 20 states. In the meantime, several states, including Vermont, Iowa, and North Carolina, have enacted "bomb-shelter" legislation, designed to protect their citizens against the more onerous aspects of UCITA. Other states are considering such legislation. Recently, NCCUSL decided not to expend further resources in support of UCITA, although it has not withdrawn the proposal.[52]

If widely adopted, UCITA would likely validate licensing arrangements that might be challenged under current law, such as whether a valid contract was formed or whether certain terms are enforceable. Overall, however, UCITA's impact would probably not be dramatic, since courts are already upholding licensing agreements involving digital media and online transactions.

Conclusion: The Uniform Computer Information Transactions Act (UCITA) is a controversial attempt to codify existing licensing law for digital media and online transactions. Widespread adoption would provide

---

(2001), *cert. denied, Taco Bell Corp. v. Wrench LLC,* 534 U.S. 1114 (2002). See also D. J. Karjala, 1997, Federal preemption of shrinkwrap and on-line licenses, *Dayton Law Review* 22: 511 (arguing that *ProCD* was wrongly decided on the preemption issues).

[52]The news release is available at <http://www.nccusl.org/nccusl/DesktopModules/NewsDisplay.aspx?ItemID=56>.

more certainty regarding the validity of restrictive licenses, although the net effect is likely to be small.

Conclusion: Although technological change has, on the one hand, made it more difficult to limit data use and redistribution, the combination of technology and contract or licensing law means, on the other hand, that data providers have the means to impose license or contract terms that limit the use or redistribution of data, a degree of control that has not been feasible for information published in paper media. In such instances, these contract or license rights have the same effects as traditional intellectual property rights, such as patent or copyright. Thus, the possibility that information providers may "lock up" geographic data must be considered and dealt with in an appropriate manner when contracting for acquisition or dissemination of geographic data and works.

5.3.1.4 Antitrust Law

Licensing arrangements are also subject to scrutiny under the antitrust laws. Potential violations include tying arrangements (in which the availability of a license for one kind of service or product in which the vendor has market power is conditioned on acceptance of a license on another service or product), a refusal to license, exclusive licenses, and blanket licenses such as those involved in the American Society of Composers, Authors, and Publishers (ASCAP), an organization that licenses its members' music.[53] Although the federal government is not subject to antitrust liability, and states and their political subdivisions generally are not liable, private actors are immune only when acting under the direction of the governmental entity. A more detailed discussion of antitrust issues is beyond the scope of this report. Additional information may be found in guidelines issued by the Department of Justice and the Federal Trade Commission.[54]

---

[53]Blanket licenses were held legal in *Broadcast Music, Inc. (BMI) v. Columbia Broadcasting System, Inc.*, 441 U.S. 1 (1979). The Supreme Court ruled that the policy of both ASCAP and BMI of offering only blanket licenses was not a per se violation of the Sherman Act.

[54]U.S. Department of Justice and Federal Trade Commission, 1995, *Antitrust Guidelines for the Licensing of Intellectual Property,* available at < http://www.usdoj.gov/atr/public/guidelines/ipguide.htm>.

Conclusion: Antitrust issues may be significant for data providers who include anticompetitive restrictions in licenses. Federal, state, and local governments generally are not liable under the antitrust laws.

5.4 FEDERAL DATA ACQUISITION AND MANAGEMENT

The decision to acquire data under license rather than by outright purchase and, if so, whether to obtain broad use and redistribution rights must be made case by case. In each instance, agencies should consider whether their missions or mandates[55] require or suggest that the data in question should be made freely available to constituents, including, where applicable, the general public.[56] This often will be the most significant step in the analysis, and agencies must exercise caution to avoid the inclination to give priority to budgetary and other considerations. They especially must avoid the temptation to construe missions narrowly so as to justify the acquisition of restricted data when the public would be better served if data were more freely available.

The second step should focus on the application of regulations affecting information policy, such as the Office of Management and Budget (OMB) circulars, the FARs, and other generally applicable regulations affecting information policy. As explained below, when agency missions do not require dissemination of, or public access to, data, we tentatively conclude that OMB Circular Nos. A-130 and A-16 permit federal agencies to acquire data subject to restrictions or limited rights. We reach a similar result concerning OMB Circular No. A-76 and the 2003 U.S. Commercial Remote Sensing Policy, concluding that although these policies may require outsourcing of data acquisition, they do not automatically mean that vendors can restrict data rights. The most specific directives on data acquisition come from the FARs, which permit the acquisition of "limited rights" data where the government does not pay the full cost of producing the data.

---

[55]We do not propose to undertake that analysis individually for the many agencies that acquire and use geographic data. Chapter 8 leads the reader through a general decision sequence for agencies.

[56]The obligation to make information public may be either implicit, where public access is required to achieve other goals, or explicit, as in the U.S. Department of Agriculture's obligation to "diffuse among people of the United States, useful information on subjects connected with agriculture." 7 U.S.C. § 2201.

Other provisions of law may require that geographic data be made available to members of the public. The accountability of agencies for their decisions, including but not limited to defending those actions in court, may require that agencies disclose information on which they relied in making policy and in taking specific actions. The Freedom of Information Act (FOIA) implements this policy by providing for any person to obtain access to agency records upon request, subject to enumerated exemptions. Licensed geographic data might be exempt from FOIA if, for example, they do not fall within FOIA's definition of records or if they constitute trade secrets or other confidential information. Acquiring data that are important to policy and decision making on terms that prevent their disclosure, however, could jeopardize an agency's ability to defend its actions. Moreover, recent changes expanding access to federally funded data and the Data Quality Act impose additional mandates for data access that may affect licensed geographic data.

5.4.1 Making Geographic Data Public

Does the law require federal agencies to acquire geographic data under conditions that allow the agency to make it available to the public?[57] The benefits of a robust public domain and ready access to public domain information have been documented in a number of reports from the National Academies,[58] and are discussed in Chapter 7, Section 7.2. Further evidence of the importance of such a policy is the Copyright Act's prohibition on the federal government's claiming copyright in the information it produces.[59] Consistent with these benefits, there are a number of statutes, regulations, and policies that point in the direction of making data, including geographic data, freely available to the public.

In 2001, the National Commission on Libraries and Information Science surveyed existing federal statutes relating to the dissemination of information[60] and identified 52 major public laws adopted between 1995 and

[57]This question is separate from whether agencies *should* acquire geographic data with limitations on the right to make the data publicly available.

[58]See, for example, NRC, 1999, *A Question of Balance: Private Rights and the Public Interest in Scientific and Technical Databases,* Washington D.C., National Academies Press.

[59]17 U.S.C. §105. Contractors that produce information for the government may claim copyright protection, however.

[60]See *A Comprehensive Assessment of Public Information Dissemination*, Vol. 1, at p. xiv, available at <http://www.nclis.gov/govt/assess/ assess.vol1.pdf>.

2000 that require federal agencies to collect and disseminate information to specialized audiences or the public.[61] Of course, there were more already on the books, and a comprehensive analysis of so many statutes is not possible in this report. Instead, this section surveys the most important authorities.

5.4.1.1 OMB Circular No. A-130 and the Paperwork Reduction Act

Agencies frequently cite OMB Circular No. A-130,[62] which implements the Paperwork Reduction Act.[63] For all executive branch agencies, including the military,[64] Circular A-130 requires the wide dissemination[65] of government information within the federal government and to the public, referencing a broad array of benefits accruing from the dissemination of information, including the furtherance of democratic processes and scientific research. Circular A-130 also discourages "improperly restrictive" practices; for example, it directs agencies to "[a]void establishing, or permitting others to establish on their behalf, exclusive, restricted, or other distribution arrangements that interfere with the availability of information dissemination products on a timely and equitable basis." Data are to be made available at the cost of dissemination, although A-130 also

---

The report, commissioned by Congress in connection with the proposed decommissioning of the National Technical Information Service, discussed then-current requirements for information dissemination by the federal government and changes needed to further dissemination.

[61]See *Index to a Compilation of Recent Federal Statutes Pertaining to Public Information Dissemination,* Appendix 33, available at <http://www.nclis.gov/govt/assess/assess.appen33.pdf>.

[62]Available at <http://www.whitehouse.gov/omb/circulars/a130/a130trans4.html#5>.

[63]44 U.S.C. §§ 3501–3521.

[64]The report focuses on data policies of and acquisition by the civilian agencies. Although some of the authorities cited apply to the military, others such as the FARs, do not. Military acquisitions are governed by the Defense Acquisition Regulations (DARs). See <http://www.acq.osd.mil/dp/dars/dfars.html>.

[65]Circular A-130 defines *dissemination* as "the government initiated distribution of information to the public." Dissemination is an affirmative obligation, irrespective of whether information is requested by anyone, in contrast to the FOIA's obligation for agencies to make "records" available upon request. See Section 5.4.3.1.

provides for exceptions, including a general one at the discretion of the Director of OMB.

The threshold question for present purposes is whether data obtained under license constitutes "government information," thereby triggering A-130's provisions. A-130 defines *government information* as "information created, collected, processed, disseminated, or disposed of by or for the federal government." Although we know of no cases expressly addressing the point, it is arguable whether data collected by private-sector firms and licensed to government fit this definition. Furthermore, A-130 nowhere mentions licenses or licensed information. Nevertheless, the foregoing definition is quite broad. Furthermore, A-130 contains several references to data that are maintained by sources other than the government. In what follows, we assume that A-130 applies to data that are acquired through licensing.[66]

At the same time, Circular A-130 indicates that proprietary rights should be respected. For example, in the section entitled "Basic Considerations and Assumptions," A-130 recognizes that the free flow of information for scientific research is subject to "applicable national security controls and the proprietary rights of others." Additional discussion in Appendix IV recognizes that federal grantees and contractors may copyright information, although the federal government may not. That section also suggests, however, that copyright as such is not a barrier to disclosure under FOIA,[67] and by implication, that copyright would not be a barrier to dissemination under A-130. Data, whether copyrighted or not, however, might be exempt from access or dissemination if one of the other FOIA exemptions applies.

A-130 also recognizes that states and localities are important sources of information utilized by federal agencies. A-130 recognizes that federal agencies must cooperate with state and local governments in the management of information and must consider the impacts of the agencies' activities on them. In another section, A-130 directs agencies to "[e]nsure

[66]One interpretation of the scope of A-130 is that "government information" is coextensive with the definition of "records" under FOIA. A-130 uses the term "government information" in reference to FOIA requirements, and also suggests that the fees authorized by FOIA access limit the fees that can be charged for dissemination under A-130. See Section 5.4.2.1, for discussion of the scope of "records" under FOIA.

[67]Appendix IV notes that FOIA "does not provide a categorical exemption for copyrighted information," suggesting that copyrighted information may in some instances be released in response to an FOIA request.

that Federal information system requirements do not unnecessarily restrict the prerogatives of state, local and tribal governments."

Some state and local providers of geographic data and works told the committee that they wish to retain the ability to charge other potential customers for the use of their databases. Their ability to do so may be impaired by free or low-cost dissemination of data they provide to federal agencies. A-130's recognition of the interests of state and local governments, together with its explicit recognition of proprietary rights, could be construed to permit agencies to acquire geographic data and works with restrictions on the distribution to other agencies or to the public, at least where such restrictions do not violate other legal requirements, which are discussed below.

Conclusion: OMB Circular No. A-130 requires the federal government to disseminate data in which it has unlimited rights (full ownership) at no more than the cost of distribution. However, A-130 probably does not prohibit agencies from agreeing to licenses that restrict redistribution or from honoring such restrictions once they have been agreed to.

5.4.1.2 OMB Circular No. A-16 and the National Spatial Data Infrastructure

Also strongly pointing in the direction of making data freely and widely available is OMB Circular No. A-16,[68] which provides for "improvements in coordination and use of spatial data," and directs the implementation of the National Spatial Data Infrastructure.[69] Although much of the circular concerns the coordination of data acquisition and management within the federal government, it endorses sharing and dissemination of geographic data among government agencies and with "non-federal users." At other points, however, the circular discusses "efficient" and "cost-effective" collection and maintenance of data, directing agencies to build on local data where possible. Thus, the circular does not say how data are to be acquired, although the general sense is that geographic data should be widely shared with the public. Restrictions on the ability of agencies to

---

[68] Available at <http://www.whitehouse.gov/omb/circulars/a016/a016_rev.html>.

[69] Circular A-16 also indicates that it incorporates Executive Order 12906 (Coordinating Geographic Data Acquisition and Access: The National Spatial Data Infrastructure), which required agencies to "adopt a plan...establishing procedures to make geospatial data available to the public, to the extent permitted by law, current policies, and relevant OMB circulars."

share data widely therefore would run contrary to that policy, even if not specifically prohibited.

Conclusion: OMB Circular No. A-16, like many federal laws, strongly favors the public availability and dissemination of government data, which would include geographic data acquired by the government. Like A-130, however, A-16 also recognizes proprietary rights and does not require that data be acquired with unrestricted rights.

5.4.1.3 Federal Activities Inventory Reform (FAIR) Act and OMB Circular No. A-76

A somewhat different conclusion may be suggested by OMB Circular No. A-76, recently revised. Circular A-76 implements the FAIR Act,[70] and requires agencies to justify engaging in "commercial" activities. Activities are presumed to be commercial unless they can be shown to be "inherently governmental." Inherently governmental activities are those that require "substantial discretion," according to A-76.[71]

A-76 could be construed as requiring outsourcing of acquisition or dissemination of geographic data, except where a particularized justification can be made for doing otherwise. However, even if A-76 and the FAIR Act require outsourcing of data requirements in some instances, that requirement does not dictate the conditions of such arrangements, such as use and redistribution rights. The reading of A-76 most consistent with other statutory and regulatory directives is that when A-76 requires an agency to outsource the acquisition of geographic data, the contract may provide for either restricted or unrestricted rights in the data. That determination is governed by other legal requirements.[72]

---

[70]Pub. L. No. 105-270.

[71]This definition has become a point of contention in litigation by two government employee unions challenging the revisions to A-76. The FAIR Act defines *inherently governmental* as those activities requiring the exercise of "discretion," an easier standard to meet than "substantial discretion" as set forth in the circular. See Federal union files suit asking court to declare revised OMB Circular A-76 illegal, *U.S. Law Week* (BNA) 71: 2829 (July 1, 2003); AFGE becomes second federal worker union to file lawsuit challenging A-76 revisions, *U.S. Law Week* (BNA) 72: 2032 (July 15, 2003).

[72]Similarly, the Commercial Space Act of 1998 requires the Administrator of the National Aeronautics and Space Administration (NASA), when consistent with scientific requirements and other conditions, to acquire "space science data" from a commercial provider. 42 U.S.C. § 14,713. However, this section also

5.4.1.4 Commercial Remote Sensing Policy

On April 25, 2003, the Office of the President announced its "U.S. Commercial Remote Sensing Policy." The policy deals with a number of topics not directly of interest in this report, but one relevant feature is a mandate to support a "robust U.S. commercial remote sensing industry" to "augment and potentially replace" some government functions and contribute to military and intelligence objectives.[73] In furtherance of this and other goals, the policy directs federal agencies to "utilize U.S. commercial remote sensing space capabilities to meet imagery and geospatial needs."[74] Similar directives apply to military and intelligence services.

Like OMB Circular No. A-76, the policy requires outsourcing of remote-sensing data collection where feasible. Similarly, the policy does not dictate the terms on which remote-sensing data can be acquired. Thus, it does not appear to dictate whether data should be acquired with unrestricted rights, or can be licensed on a more limited basis.

Conclusion: OMB Circular No. A-76 and the Commercial Remote Sensing Policy require outsourcing by the federal government of all functions that are not "inherently governmental." Neither policy, however, specifies the terms on which data should be acquired or when broad use and redistribution rights should be obtained. Thus, policies that otherwise encourage or require the dissemination of geographic data need not be affected by outsourcing requirements. This issue has not been legally tested, however.

5.4.1.5 FARs

Agency acquisitions of data, as with their acquisitions of other goods and services, are governed by FARs, both those that are generally applicable, which are discussed below, and those adopted by specific agencies.[75]

---

states that "[n]othing in this subsection shall be construed to preclude the United States from acquiring, through contracts with commercial providers, sufficient rights in data to meet the needs of the scientific and educational community or the needs of other government activities."

[73]Fact Sheet, U.S. Commercial Remote Sensing Policy § III (Apr. 25, 2003). The fact sheet indicates that the policy is not intended to have the force of law.

[74]*Id.* at p. 4.

[75]Provisions dealing with the acquisition of data are found in 48 C.F.R. Parts 27 and 52. Department of Commerce Acquisition Regulations contain no additional

In general, FARs specify contract terms in more detail than any of the statutes and regulations discussed earlier.[76]

At the outset, the regulations recognize that agencies may need to acquire data for many different purposes.[77] Perhaps for this reason, agencies are allowed a great deal of latitude in how they contract for data acquisition. Contracts for data must provide for the respective rights of the parties in the data acquired.[78] The regulations set forth policies for the acquisition of data and permit agencies to adopt alternative policies and contract clauses only to the extent necessary to meet the requirements of other laws.[79]

The FARs make a fundamental distinction between data produced under federal contracts and data gathered at private expense. Thus, the FARs give the government "unlimited rights" in data "first produced in the performance of" the contract, while recognizing the contractor's right to copyright scientific articles containing or based on data first produced under the contract or with the agency's permission.[80] In contrast, *limited rights data* is defined as data developed through private expense, implying that data developed through government funding should not be

---

provisions pertaining to the acquisition of data. Further analysis of agency-specific acquisition regulations is beyond the scope of this discussion. 48 C.F.R. Parts 27 and 52 are undergoing revision at this writing. See Federal Acquisition Regulation; FAR Part 27 rewrite in plain language, 68 Fed. Reg. 31,790 (May 28, 2003).

[76]The Brooks Architect-Engineers Act (the "Brooks Act"), Pub. L. No. 92-582, provides for qualification-based, negotiated contracting for the services of architects and engineers, which has been interpreted to include some mapping and surveying services (48 C.F.R. § 36.601-4). Recently, the Department of Defense (DoD), the General Services Administration (GSA), and NASA solicited comments on the scope of the Brooks Act's application to mapping (DoD/GSA/NASA, Federal acquisition regulations; application of the Brooks Act to mapping services, request for comments, 69 Fed. Reg. 13,499 [Mar. 23, 2004]). Brooks Act contracts are nonetheless subject to the generally applicable contracting requirements of the FARs, however, except where the general requirements conflict with the specific Brooks Act regulations of 48 C.F.R. Part 36.

[77]48 C.F.R. § 27.402. For a general discussion, see W. N. Keyes, 1996, *Government Contracts Under the Federal Acquisition Regulation* §§ 27.21–27.26, Eagan, West Information Publishing Group.

[78]48 C.F.R. § 27.403. This section directs the agency to use the data rights clauses provided in section 52.227-14, which has five alternative versions, "where determined to be appropriate," but also allows other versions to be used.

[79]48 C.F.R. § 27.101.

[80]48 C.F.R. § 27.404(f).

subject to limited rights.[81] Variation from this standard form, however, is permitted to allow the government to obtain limited-rights data and to limit its uses of the data to those specified, particularly when the data are obtained for particular purposes not inconsistent with restricted uses.[82]

The provisions governing cooperative research and development activities provide even more latitude, and do not recommend or require specific clauses.[83] This regulation also makes clear, however, that the government's rights should be limited only in the event, and to the extent, that the contractor makes a substantial contribution of its own resources in developing the data.[84]

Conclusion: The FARs specify clauses concerning data rights to be included in data acquisition contracts and require agencies to acquire unlimited rights in data developed at government expense.

Conclusion: Under the FARs, restrictions on a government agency's right to use or distribute data are appropriate when the government is not compensating the vendor for all of the costs of producing the data (as when the government acquires a nonexclusive right to use preexisting geographic data or when the government contracts to pay only a portion of the cost of acquiring new data).

5.4.2 Other Legal Requirements for the Disclosure of Data

In addition to the foregoing authorities that govern agencies' data management practices generally, there are a number of statutes and regulations that require the government to disclose data in particular circumstances. These provisions usually apply to geographic data. Several of the more prominent ones are summarized below.

---

[81]With respect to data not first produced under the contract, the contractor is prohibited from delivering such data pursuant to the contract unless the contractor provides the government with a license to use the data on the same terms as provided for data produced under the contract. The contractor is permitted to retain proprietary rights in computer software, however, and to license software to the government. In that case, the government obtains "restricted rights" to the software. 48 C.F.R. § 52.227-14.

[82]48 C.F.R. §§ 27.404, 27.405.

[83]48 C.F.R. § 27.408.

[84]*Id.*

5.4.2.1 FOIA

The FOIA[85] requires federal agencies to disclose upon request, records that they have created or maintained, unless the records are covered by one of the exemptions of the Act.[86] Records are not limited to paper copies, and include information stored in any form, including electronic.

A threshold question is whether licensed data constitute "records" subject to FOIA.[87] At least one court has held that a database licensed to a federal agency under conditions that "greatly restrict[ed]" the agency's control did not constitute records, and therefore did not fall within the ambit of FOIA.[88] Arguably, this ruling would apply to licensed geographic databases, but the paucity of legal rulings on the issue leaves open the possibility that data subject to a restrictive license would fall within the scope of FOIA records. We assume that result for the purposes of the ensuing discussion.

Even if licensed data constitute FOIA "records," however, they may be protected from disclosure by one of the statute's multiple exemptions. Several exemptions are potentially important in the context of this report. There is a specific exemption for geological and geophysical information and data concerning wells.[89] A broader exemption applies to trade secrets, and commercial or financial information of a privileged or confidential nature.[90] Geographic data provided by private vendors could qualify for this exemption, if the information has been maintained as confidential, is

---

[85]5 U.S.C. § 552. See also 1 R. J. Pierce, Jr., *Administrative Law Treatise* §§ 5.2–5.16 (2002) (hereinafter 1 *Administrative Law Treatise*).

[86]5 U.S.C. § 552(b).

[87]See discussion of the scope of "government information" under OMB Circular A-130, in Section 5.4.1.1.

[88]*Tax Analysts v. U.S. Dept. of Justice,* 913 F. Supp. 599 (D.D.C. 1996). In *Gilmore v. Dept. of Energy,* 4 F. Supp. 2d 912 (N.D. Cal. 1998), the court held that software licensed to the government did not constitute records subject to FOIA because the agency had only a license to use the software and, alternatively, because the software and related technical information did not "illuminate the structure, operation, or decision-making structure" of the agency. A third ground for the result was that the software qualified as "trade secrets or commercial or financial information" that was confidential or privileged.

[89]5 U.S.C. § 552(b)(9). This exemption seems to be concerned with oil and gas wells and the potential harm to the competitive positions of participants in the oil and gas industry. 1 *Administrative Law Treatise* § 5.16; 2-1. J. A. Stein, G. A. Mitchell, and B. J. Menzines *Administrative Law* § 10.01 (2003).

[90]5 U.S.C. § 552(b)(4). See 1 *Administrative Law Treatise* § 5.10; 2-10; 2-1 *Administrative Law* § 10.05.

used in their business, and is licensed to the government with distribution restrictions. Geographic data licensed from state or local governments also might qualify for this exemption if the data or database have been maintained in confidence.[91] This exemption is evaluated under a balancing test, however, in which the court weighs the public interest in understanding the operations of government against the interest Congress intended the exemption to protect—among other things, the harm to the competitive position of the information provider and potential harm to the government's ability to acquire similar information in the future.[92]

In most instances, it is reasonable to expect that licensed data would not be subject to disclosure under FOIA because it will fall outside the scope of FOIA or will enjoy the benefit of one of its exemptions. Where applicable, licensors of geographic data may wish to state in the license that the licensed information is trade secret or confidential. The courts ultimately have the authority to determine whether contested information meets the requirements of the definition, however, and the party arguing for nondisclosure (agency or private party) has the burden of proof that the requirements of an exemption are met.

Conclusion: Licensed data ordinarily will not be subject to disclosure under FOIA, either because they do not constitute agency records or because they qualify for FOIA's exemption for trade secrets or other confidential commercial information. Parties to a data license should state their understanding that the information falls within one of the exemptions, but any such designation may be subject to subsequent judicial review.

5.4.2.2 Accountability and Judicial Review of Agency Actions

Another constraint on agencies' ability to limit public access to commercial geographic data or works is the need for public access to the rationale of certain agency actions and policy decisions, such as rulemaking. Statutes that may come into play, in addition to FOIA, include the

---

[91]This exemption also requires that the information have been obtained from a "person" to qualify for the exemption; a Department of Justice memorandum indicates that state governments (and presumably their subdivisions) are "persons" within the meaning of the statute, although it has been interpreted to exclude federal agencies (*Administrative Law*, see footnote 89; Department of Justice Guide to the FOIA, 2002, Appendix 10A).

[92]*Gilmore,* 4 F. Supp. 2d at 922.

Government in the Sunshine Act,[93] the Federal Advisory Committee Act,[94] and other provisions of the Administrative Procedure Act. Agencies usually are required to explain rules and other actions they propose by publication in the *Federal Register*, giving the public access to the agency record and an opportunity to comment.[95] These decisions typically are subject to judicial review,[96] which often requires examination of the factual and legal bases of the decision.[97] These kinds of decisions could involve geographic data.

Assume, for example, that the designation of a critical habitat under the Endangered Species Act is based in part on mapping of a watershed. If the decision is challenged in the courts on the basis that the mapping was inaccurate, the agency must be able to point to supporting information in the administrative record, and that information ordinarily must be available for public scrutiny during the regulatory process and any challenges brought.[98] If the agency were unable to produce the information supporting its decision because of contractual restrictions, its decision could be overturned. The cases are not entirely clear as to how strictly this requirement would be applied, however, and the answer may depend on how central the information is to the decision under review.[99]

Conclusion: An agency must be able to disclose geographic data that it relied on in policy decisions and other actions, including rulemaking. If the agency is unable to make such data available because of contractual restrictions, the agency's action may be overturned.

[93]5 U.S.C. § 552(b).

[94]5 U.S.C. Appendix I.

[95]1 C. H. Koch, Jr., *Administrative Law & Practice* §§ 4.10–4.18, 4.30–4.34 (2d ed. 1997).

[96]2 *Administrative Law & Practice* § 8.23 (2d ed. 1997).

[97]2 *Administrative Law & Practice* § 8.27 (2d ed. 1997).

[98]1 *Administrative Law & Practice* §§ 4.32, 4.44 (2d ed. 1997).

[99]In *National Nutritional Foods Ass'n v. Mathews,* 418 F. Supp. 394 (S.D.N.Y. 1976), *rev'd on other grounds,* 557 F.2d 325 (2d Cir. 1977), the court upheld an action by the Food and Drug Administration despite the fact that the FDA relied in part on information that it withheld from public review under FOIA. Thus, FOIA may trump the requirement of public access to the record, although if the information were important to the decision, the decision might be reversed or remanded. See also *Mortgage Investors Corp. v. Gober,* 220 F.2d 1375 (Fed. Cir. 2000) (rule upheld despite agency's withholding of some information it relied on).

5.4.2.3 Data Access and the Data Quality Act (DQA)

Two recent enactments are also designed to strengthen public access to information that government agencies use in formulating rules and policy. In 1999, Congress adopted the Shelby Amendment to an appropriations bill. The amendment required OMB to amend Circular No. A-110, to "require Federal awarding agencies to ensure that all data produced under an award will be made available to the public through the procedures established under the Freedom of Information Act." The revisions to Circular A-110 limit access under FOIA to "research data relating to published research findings produced under an award that were used by the Federal Government in developing an agency action that has the force and effect of law."[100]

The significance of this data access amendment is that it subjects data in the hands of federal grantees to FOIA, which previously extended only to information in the possession of the federal government. This includes, for example, the data produced by academic researchers pursuant to federal research grants when the data are cited in support of a federal regulation.[101] Like FOIA generally, A-110 protects trade secret information, but a recent request for access to proprietary software suggests that pressure for access to otherwise nonpublic information is likely to increase.[102]

The DQA[103] followed closely on the heels of the Shelby Amendment in the fiscal 2001 appropriations bill. It directs OMB to issue guidelines that "provide policy and procedural guidance to Federal agencies for ensuring and maximizing the quality, objectivity, utility, and integrity of information (including statistical information) disseminated by Federal agencies." The OMB, in turn, required agencies to issue their own guidelines. Those guidelines must include "administrative mechanisms" for "affected persons" to challenge the quality of information disseminated

---

[100]OMB Circular No. A-110, available at <http://www.whitehouse.gov/omb/circulars/a110/a110.html>.

[101]More specific conditions for access are set forth in Circular A-110.

[102]On January 22, 2002, the Center for Regulatory Effectiveness, a group instrumental in the passage of The Shelby Amendment and the DQA, sent a letter to the U.S. Environmental Protection Agency (EPA), requesting that EPA obtain the rights to release proprietary software that the agency is using to predict economic effects of "Multi-Pollutant" air pollution. The letter is available at <http://www.thecre.com/quality/20020121_cioletter.html>.

[103]Section 515 of the Treasury and General Government Appropriations Act for Fiscal Year 2001 (Pub. L. No. 106-554).

by the government and for "correcting" information that does not meet the guidelines.

In contrast to Circular A-110, which concerns access to research data in the hands of grantees, the DQA is directed to information disseminated by the government. This distinction may be more apparent than real, however, since the DQA can be used to challenge scientific conclusions disseminated by the government, but based on academic or private research. Recent actions include challenges to the National Assessment on Climate Change[104] and to research on the herbicide atrazine.[105] Although the standards for "correction" of information under the DQA have not been developed,[106] they will almost surely require some examination of underlying data. It would not be surprising for geographic data to be implicated in DQA challenges and subject to public disclosure in the process. Alternatively, without the ability to disclose data, information and decisions that cannot be justified may have to be withdrawn.

Conclusion: Newly adopted data quality and data access requirements may necessitate the disclosure of geographic data, particularly where the data form the basis of a government policy or regulation. The scope of these requirements is uncertain, however.

5.5 STATE AND LOCAL LAW AND POLICY

State and local governments generate and collect significant quantities of geographic data utilized by federal programs, and they are also consumers of geographic data supplied by the federal government and other entities. Federal law permits state and local governments to assert copyright in works containing geographic data (if they otherwise meet the requirements for copyright protection). When consistent with local law, state and local governments may also maintain geographic data as secret, or to restrict their use and redistribution. Thus, state and local law or government policies may place important conditions on how geographic data are obtained from and delivered to states, counties, and municipalities.

---

[104]See the Center for Regulatory Effectiveness discussion at <http://www.thecre.com/access/index.html>.

[105]The Center for Regulatory Effectiveness also brought this challenge. See <http://www.thecre.com/quality/index.html>.

[106]Many issues on the applicability and scope of the DQA are in dispute and have not been resolved by the courts.

There are two major issues on which state and local law or policy is likely to affect transactions in geographic data and works.[107] First, these entities sometimes rely on cost recovery to fund their collection and related activities (Chapter 4, Section 4.3). In these instances, they are likely to anticipate multiple licenses of a dataset and are therefore unwilling to sell data outright or license it with permission for the federal agency to distribute it publicly. Thus, where cost recovery policies are in place, it may not be possible for federal agencies to obtain unlimited rights to geographic data.

Moreover, federal policies requiring the disclosure of geographic data may inhibit state and local participation in partnering with the federal government to generate databases such as *The National Map*. For example, Executive Order 12906, [108] encourages partnerships between federal agencies and state, local, and tribal governments to share costs in acquiring geographic data. Because the order also requires public access to the data,[109] state and local governments may find such partnerships to be contrary to their interests or their laws.

State open records laws sometimes make the cost recovery policies discussed above somewhat problematic, however, because these laws require disclosure of public records upon request of citizens, much like the federal FOIA.[110] Pursuant to these statutes, some state courts have required the disclosure of geographic datasets developed in a municipality or county, effectively putting the information in the public domain.[111] Recognizing

[107]See also discussion in Section 5.3 on contract law principles, which are also a matter of state law.

[108]Coordinating geographic data acquisition and access: The National Spatial Data Infrastructure, 59 Fed. Reg. 17,671 (Apr. 13, 1994), *amended by* Executive Order 13286 of February 28, 2003, Amendment of executive orders, and other actions, in connection with the transfer of certain functions to the Secretary of Homeland Security, 68 Fed. Reg. 10,619 (Mar. 5, 2003). Executive Order 12906 has been incorporated into OMB Circular No. A-16. *See* footnote 69.

[109]Access is required "to the extent permitted by law, current policies, and relevant OMB circulars, including OMB Circular No. A-130…and any implementing bulletins." *See* 59 Fed. Reg. 17,671, cited in footnote 108.

[110]See Conn. Stat. Ann. § 1-210. For a compilation, see Urban and Regional Information Systems Association (URISA), 1993, *Marketing Government Geographic Information: Issues and Guidelines,* Washington, D.C., URISA, pp. 12–22.

[111]In some instances, state opens records laws have been interpreted to permit access and copying, but not redistribution or other uses, such as commercial uses, if the locality prohibits those activities or requires a license for them. *See, for example, County of Suffolk, NY v. First American Real Estate Solutions,* 261 F.3d 179 (2d Cir. 2001), in which the county sued First American for copyright

the difficulties that these disclosure requirements pose for the prospect of cost recovery, several state legislatures are considering legislation to permit their political subdivisions to restrict access to or copying of geographic data.[112]

Conclusion: State and local governments are both suppliers and consumers of geographic data. Not uncommonly, they rely on revenue from licensing their geographic data to recover some of their costs, which may limit federal agencies' ability to acquire unlimited rights in the data. The public disclosure requirements of federal law may inhibit state and local participation in partnerships to acquire geographic data.

5.6 SUMMARY

Licensing of geographic data and works has come of age because of the limitations of copyright and other intellectual property doctrines in protecting them in the digital environment. Copyright protection is often unavailable for geographic data and is limited for databases and datasets of geographic data. Even remote-sensing imagery and maps are likely to enjoy only limited protection. The extent to which copyright applies to particular geographic data and works is often uncertain.

With limited copyright protection, providers of geographic data or works in digital form have turned to other means to protect these works. These include using technological means to control access and copying,

---

infringement of its official tax maps. This case involved the application of New York's Freedom of Information Law (FOIL), which permitted inspection and copying. The court concluded that the FOIL did not abrogate copyright protection, which in this case was invoked to prohibit commercial redistribution of the copyrighted maps. *See also* Lawsuit against property appraiser could set precedent in GIS cases, *Naples Daily News* (Mar. 28, 2004) (summary judgment for Collier County on its right to restrict commercial use of geographic information system data through licensing).

[112]See, for example, Connecticut H.B. 5014 (2003). The proposed bill exempted from disclosure "municipal geographic information system data concerning a residence or building," but a subsequent version of the bill prohibited the sale of certain geographic information paid for with public funds. Hawaii H.B. 443, deferred to the 2004 session, would delete "any map, plan, diagram, photograph, photostat, or geographic information system digital data file" from the definition of government records that the government is required to provide copies pursuant to its open records law. See <http://www.capitol.hawaii.gov/sessioncurrent/bills/hb443_.htm>.

measures that are reinforced by the DMCA for works having at least some copyright protection. Moreover, courts recently have upheld contracts or licenses that limit the uses that a licensee can make of data, or that prohibit further distribution, rejecting arguments that shrink-wrap and click-wrap licenses are not agreed to by the consumer or that the contractual protection of facts is preempted by copyright law. Data providers' rights are likely to be further strengthened if Congress adopts database protection.

Government agencies face difficulties in adapting their data acquisition policies to this new, changing environment. Uncertainty can be reduced by thoughtful drafting; and contracts for the purchase or licensing of data should address the rights conveyed and withheld, including whether copyright is claimed, rights are assigned or licensed, and rights are withheld—such as uses that can be made of the data or works, and the persons or entities authorized to use it. Additionally, government agencies acquiring data subject to limited rights should consider their technical capability to manage the restrictions, as well as other costs associated with managing geographic data in which they have limited rights.

Federal agency data acquisitions also are constrained by the requirements of a variety of federal laws and regulations. Some federal laws and policies embody a strong preference for making data available to the public and a number of the agencies told the committee that their missions require them to acquire unrestricted geographic data that are free to make available to the public on the Internet. Additionally, government accountability may require that geographic data be available to the public, particularly under changes to the law regarding data access and data quality. Even so, OMB Circulars such as A-130, the FARs, and the FOIA recognize the possibility that some government information will be subject to proprietary restrictions and cannot be disseminated or made available to the public.

VIGNETTE E. A PERSONAL COMMUNICATOR DREAM

Susan carries a pocket-sized personal communicator that receives, sends, and records voice communications, text messages, still images, and videos. Using voice or keypad commands, she can connect to the Web, comparison shop, or make mail-order purchases at any time from any location. The communicator gives Susan information about any building or commercial establishment at which the device is pointed, provides voice directions or route maps for any prescribed destination, tracks her as she travels and reroutes her around traffic congestion, allows her to communicate simultaneously with multiple friends by oral conferencing or text messaging, and provides her with a range of "pull" services that answer such questions as whether stores in her vicinity sell aspirin. At any time, Susan can set or change her privacy preferences dictating who can contact her through the device and by what methods, whether and how precisely in time and location they may track her current and past positions, and to what extent the telecommunications provider may archive her past locations, purchases, and activities accomplished through the device.

Susan's personal communicator contains all of her work and personal documents from the past 10 years, several movies and games, thousands of songs, and high-resolution images of Earth's land masses with sufficient detail to do realistic aerial or ground-level flythroughs down any street worldwide or the hallways and rooms of any public building. Although public officials in Susan's town use these flythrough capabilities on their personal communicators to manage facilities and to provide evacuation paths through buildings or along street networks during emergencies, Susan and her friends use such flythroughs to play virtual games and explore real-world settings where they have never physically been. When updates containing more detailed or more comprehensive geographic data become available, Susan simply downloads the upgrade for a fee from the vendor or for free from the information commons.

Millions of individuals like Susan also carry personal communicators, and each person regularly makes decisions about (1) the ways others can communicate with them, (2) what information about them will be available to others, and (3) what digital services and products they are interested in. Underpinning this environment is a network of legally enforceable contracts and licenses controlled by each person's preference settings and enforced automatically through computer code.

In the end, the dream comes down to this: Can the core ethical principle of individual autonomy be used to direct the creation of an overall information infrastructure that automatically enforces contracts and licenses to efficiently support an active information commons and a thriving marketplace?

6

Economic Analysis

6.1 INTRODUCTION

This chapter summarizes what economic theory has to say about how the terms on which government obtains data from, or provides data to, users—including through licensing or purchase—affect the well-being of society. The chapter is divided into five sections: first it describes the factors that affect whether data are likely to be produced and distributed efficiently; second, it emphasizes that licensing is just one of several options for government data procurement, and explains why some options are better suited to particular agency missions than others; third, it examines the advantages and disadvantages of licensing as a tool for achieving economic efficiency; fourth, it describes the economic considerations when agencies decide to negotiate licenses; and fifth, it discusses how agency decisions to charge a user fee may promote or hinder economic efficiency.

6.2 ACHIEVING THE GOAL OF ECONOMIC EFFICIENCY

The first step in any economic analysis is to specify goals: What is society trying to accomplish? Like most commentators, we identify *economic efficiency* as an important goal. In everyday life, economic efficiency has many different meanings. To an economist, however, economic efficiency *requires that there remain no unexploited opportunities to make someone better off without making someone else worse off.*[1] In the context of the provision of information,[2] this subsumes two criteria. First, *efficiency in production* requires that society create information if and only if its cost is less than its combined value to all users. Second, *efficiency in distribution* requires that information be available to all users who value it at or above the marginal cost of distribution.[3] For reasons discussed below, it is rarely possible to completely achieve either, let alone both, of these objectives. Nonetheless, society has developed a number of institutions that move it closer to their achievement. Where benefits are not quantifiable, one may have to be satisfied with achieving the agency's goal (quantitative or not) at minimum cost.

Unlike most commodities, information goods such as geographic data have the property that once generated they can be used by one user without reducing the amount available for use by others. As previously noted, efficiency in production requires that society produce information if and only if the information's *combined* value to all users exceeds the cost of production. For example, if user A values a particular piece of information at $5 and user B values the same information at $3, it is economically efficient to produce the information at any cost less than $8. For many types of information, the value to one user is independent of the number of other users. In these cases, the information's value to society is simply the sum of its value to each individual user.[4]

---

[1]An alternative definition of efficiency is the state in which all scarce resources are directed to their most highly valued uses.

[2]We use the term *information* rather than *intellectual property* because whether information becomes intellectual property depends on whether the legal regime accords it that treatment. For example, information that is in the public domain is not intellectual property.

[3]*Marginal cost* refers to the cost of providing a copy to an *additional* user. This should be distinguished from the *average cost* of making copies for *all* users. "Cost of dissemination" (terminology in Office of Management and Budget [OMB] Circular No. A-130), and "direct cost of search, duplication, and review" (Freedom of Information Act [FOIA] terminology) usually set prices close to the economic concept of marginal cost, but differences may exist in particular circumstances. Unless otherwise noted, this report assumes that FOIA and OMB Circular No. A-130 require marginal cost pricing, and, for convenience, we adopt the phrase "marginal cost of distribution" to represent any of the legally or economically precise terms listed in the preceding sentence.

[4]However, there may be cases in which the value of information to any user is reduced when others have access to the same information. In such cases, the social value of the information will be less than the sum of the values to each exclusive user.

Once information has been produced, economic efficiency requires that it be made available to any user who values it at or above the cost of dissemination. This condition can be met as long as the price charged to any individual is *no greater than* the amount that individual would be willing to pay for the information but *no less than* the marginal cost of distribution. Where the price exceeds the cost of distribution, however, the number of consumers who acquire and use the information is likely to be inefficiently reduced. When this occurs, the result is a deadweight loss because some members of society lose benefits without there being any gains to others.

It is also important to recognize that users of geographic data include not only final *consumers* but also other *producers* of information products or services, so that efficient distribution is likely to contribute to efficient production.[5] In particular, the ability of other vendors, academic researchers, and government agencies to use geographic data in their own research and development activities may substantially enhance their incentives to invest in the creation of valuable new geographic products and services. Unless the conditions for efficient distribution are met, therefore, not only will some *consumers* fail to obtain information that they value at more than its incremental cost but the costs for some *producers* also will be increased. Because the production of information goods tends to be cumulative, with one creator building on the efforts of others, inefficient distribution may therefore raise the cost of subsequent research and development.[6] Conversely, later innovators have lower costs when they can build freely on the work of their predecessors.[7] Fostering the creation and development of geographic data and services requires striking the right balance between proprietary rights and free or open access to information.

---

[5]See examples of the impacts of contrasting dissemination regimes on production described in Chapter 4, Section 4.3.

[6]The costs to later developers can be reduced either by limiting the rights of earlier ones or by lowering the costs of transactions between them. In Chapter 9, we discuss mechanisms that might lower transaction costs.

[7]The impact of intellectual property protection on the costs of innovation is discussed, for example, by W. M. Landes and R. A. Posner, 1989, An economic analysis of copyright law, *Journal of Legal Studies* 18: 325–363; and S. Scotchmer, 1991, Standing on the shoulders of giants: Cumulative research and the patent law, *Journal of Economic Perspectives* 5: 29–41. For a different view, see E. W. Kitch, 1977, The nature and function of the patent system, *Journal of Law and Economics* 20: 265–290.

Even if it were possible to achieve *either* efficiency in production *or* efficiency in distribution—no small task—achieving *both* simultaneously is even more difficult.

Permitting producers to charge high prices increases their incentives to produce information. At the same time, high prices exclude some would-be consumers from the market.[8] Thus, the benefits of encouraging additional innovative activity generally must be balanced against the social costs of discouraging efficiency in distribution. Although high prices may be necessary to encourage the production of information, high prices create inefficiencies by excluding users who are willing to pay at least the cost of distribution but not the prices being charged by producers.

6.2.1 Other Sources of Inefficiencies in Information Markets

The deadweight loss that results when the price of information exceeds the cost of distribution is not the only type of inefficiency that can occur in a market for information. This section describes six additional inefficiencies that also can contribute to deadweight loss.

First, information is an "experience good." This means that a potential buyer may be unwilling to pay for a particular piece of information before inspecting it because the buyer cannot place a value on it. On the other hand, the buyer also may be unwilling to pay *afterward* if inspection reveals all of the needed information.[9] This problem can be overcome in cases where a small subset of data is enough to demonstrate quality. Such is typically the case in the geographic data market. Furthermore, users who repeatedly sample without buying eventually will be denied the opportunity to sample, and suppliers who attempt to charge high prices for information of limited value eventually will find that their customers go elsewhere. However, sampling and the use of reputation

---

[8]However, as noted earlier, it also may increase the *costs* of producing information.

[9]This is described by Kenneth Arrow (K. J. Arrow, 1962, Economic welfare and the allocation of resources for invention, in R. R. Nelson, ed., *The Rate and Direction of Inventive Activity: Economic and Social Factors*, National Bureau of Economic Research Conference Series, Princeton, NJ, Princeton University Press, pp. 609–625): "There is a fundamental paradox in the determination of demand for information; its value for the purchaser is not known until he has the information, but then he has in effect acquired it without cost." Of course, providing a "sample" may be sufficient to reveal the value of the information without completely eliminating the incentive to pay for it.

may be unable to eliminate completely the inefficiency that arises from this source.

Second, there may be strong incentives for users to underreport the value of information in order to obtain lower prices from the producer. If a large proportion of users succeed in this strategy, however, the information may not be created at all or the amount may be inefficiently small. In practice, high-volume users often accept the fact that, because small-volume users will "free ride," they must pay a disproportionate share of the cost of producing information. As long as the contributions of these users are sufficient to cover these costs, the information will be produced, although perhaps less will be produced than if other users had also paid to do so.

Third, when different users place different values on the information, vendors may have to charge different prices to each user, that is, practice price discrimination. This explains why some vendors have attempted to ascertain how individual users will employ their data.[10] In the case described earlier, the only *single* price at which both A and B will purchase the information is $3. However, if the cost of producing the information exceeds $6 and price discrimination is impossible, no private entity would supply it despite the fact that it would be efficient to do so as long as the cost of production is less than $8. Even though users may try to disguise the true value of the information to them, licensors sometimes can use objective criteria to identify consumers willing to pay higher prices.[11] Nonetheless, effecting price discrimination is generally quite difficult and, even where it is accomplished, it is unlikely to satisfy fully the conditions for efficient production and dissemination of information.

Fourth, it may be hard to exclude nonpaying users, especially when paying customers choose to share information. That is, it may be impossible for producers to prevent free riding once their information goods have been disseminated to a large number of users. Vendors can respond by (1) suing those who engage in or facilitate such behavior, (2) using technical protections such as providing software in copy-protected form or requiring that playback devices be designed to limit copying, (3)

---

[10]For example, see footnote 44 of Chapter 3.

[11]For example, academic journal publishers typically impose higher subscription rates on libraries than on individuals. Similarly, music-performing license organizations impose higher fees for performing rights on establishments such as nightclubs and bars with large revenues than on the same types of establishments where revenues are smaller. Finally, book publishers charge higher prices for hardbacks than for paperbacks, presumably because those who wish to obtain a book as soon as it is published are willing to pay higher prices.

obtaining indirect compensation by imposing levies on recording media or devices,[12] (4) pricing products at low rates that discourage unauthorized copying, or (5) imposing high fees that capture some of the value that otherwise would be lost to sharing.[13] Nonetheless, some free riding is likely to occur, with the result that efficiency in the production of information may not be fully achieved.

Fifth, a special difficulty may be created when users require access to information that is owned by *different* entities in order to have anything of value, a difficulty variously known as the "complements" or "anti-commons" problem. In these circumstances, each individual owner may attempt to capture a disproportionate share of the combined value of the information. However, if all owners attempt to do this, the transaction may never occur, or occur at an inefficiently high price, so that no value is created, or the value that is created is inefficiently low.[14] Recognizing this problem, individual producers may attempt to prevent or discourage such behavior by internalizing the distribution of profits, perhaps by merging, or by forming joint ventures or consortia. Alternatively, as in the case of patents, producers may cross-license or form patent pools in which each producer agrees to make licenses to its patents available at little or no cost in return for an agreement on the part of other producers to do the same. Nonetheless, these institutions may deal only imperfectly with the complements problem.

Finally, transactions may not occur because the task of negotiating and administering contracts entails high transactions costs. Transactions involving information goods may be especially complicated (and expensive) compared to conventional sales in which the vendor transfers complete title at the time of payment.[15] Strategies for reducing transaction costs include (1) widespread adoption of standardized contracts and other business practices, (2) blanket licenses that give the licensor access to all of the works in the owner's collection in return for a fixed fee,[16]

---

[12]For example, some countries have imposed levies on CD-ROMs, with the proceeds used to compensate record companies for unauthorized copying.

[13]Professional journals often claim that they charge a high subscription rate to academic libraries because they need to capture some of the value that individual library patrons place on their contents.

[14]Chris Friel related such a case in testimony to the committee, wherein negotiations involving multiple data sources became almost insurmountably bogged down.

[15]Indeed, testimony to the committee from licensors and licensees supports this assertion.

[16]Information owners also may band together to license their works through a single organization, although this could raise antitrust concerns. Examples

and (3) creating centralized marketplaces or other institutions where buyers and sellers can easily find each other. Finally, producers might accept, or even promote, limitations on their own intellectual property protection if, in return, that also will increase their access to the information of others.

6.3 THE CHOICES: GOVERNMENT MISSIONS AND PROCUREMENT OPTIONS

Licensing is one of several tools available to agencies when they procure geographic data,[17] and consideration of all options is important in any economic analysis. The suite of workable options in each case is determined by what the agency is trying to achieve, and will be influenced by whether the agency's mission requires broad redistribution, limited redistribution, or only internal use. In practice, most of the potential policy benefits from licensing tend to involve the last two categories. Here, government agencies often can obtain useful (though limited) rights at prices that are generally below the costs of outright purchase or in-house production.[18] However, the savings may be small if there are relatively few limits on the rights that are transferred. As a result, licensing is unlikely to be chosen when an agency's mission involves acquiring data that it wishes to disseminate to the public as a whole without restriction.

6.4 WHEN IS LICENSING ECONOMICALLY EFFICIENT?

In this section we weigh the advantages and disadvantages to a government agency of obtaining data from the private sector through

---

include the American Society of Composers, Authors, and Publishers and Broadcast Music, Inc. (performance rights for music) and the Copyright Clearance Center (reproduction rights). For a (somewhat dated) description of these organizations, see S. M. Besen and S. N. Kirby, 1989, *Compensating Creators of Intellectual Property: Collectives That Collect*, R-3751-MF, Santa Monica, The Rand Corp. See also J. H. Reichman and P. Samuelson, 1997, Intellectual property rights in data? *Vanderbilt Law Review* 50: 51.

[17]See Chapter 4.

[18]Recall that we have defined purchase to mean that there are no limits on the rights of the acquirer to use or transfer data.

licensing, as opposed to purchasing the data from the private sector or producing the data in-house, using economic efficiency as our criterion.

6.4.1 Economic Advantages of Licensing

Licensing data from the private sector is likely to be economically efficient when government does not have enough information to decide what data society needs. In this case, entrepreneurs may have better or different information about which investments make sense. Private markets are a powerful mechanism for eliciting and organizing information held by large numbers of scattered actors, in which case government may be just one of a large number of customers for the data that are produced.

When private firms have superior information about potential demand for data, it is likely to be economically desirable to decentralize the investment decision to dozens and perhaps hundreds of private-sector entrepreneurs.[19] Moreover, large specialized private-sector data producers may benefit from economies of scale and scope that might not be available to individual government agencies for which data acquisition are a small part of their missions. Markets also guard against the possibility that government-funded programs will take on a life of their own; forcing actors to invest their own money eliminates the danger of program managers, employees, or outside contractors acquiring a personal stake in continuing unproductive programs. This problem is likely to be reduced, however, when society's need for a particular dataset enjoys widespread consensus (e.g., data for basic navigation or emergency response), in which case outright purchase or in-house production may be preferred.

The foregoing analysis is quite general, and there may be additional economic efficiencies from acquiring data with licensing restrictions in specific circumstances. We provide four examples.

First, acquiring data through outright purchase or in-house production in order to avoid restrictions on use may not be efficient if the number of potential users is small. In such cases, users who are most directly affected are likely to press an agency to be able to distribute the data freely even when the benefits they receive are smaller than the costs that the agency incurs in being able to do so. An important corollary is that, from a purely logical standpoint, agencies should fund data acquisition at the smallest

[19]Firms may also have superior information about the technical feasibility of creating a particular product. However, such information is usually well known for geographic data.

level of government that embraces all potential beneficiaries,[20] although it may be difficult to apply this principle when, as is often the case, the potential beneficiaries cannot be identified in advance.

Second, agencies may find it efficient to accept licensing restrictions that allow them to "piggyback" data collection or distribution activities onto commercial programs even where the result is that the agencies do not acquire the precise data that they need. One example of the synergies is the NEXTMap Britain project, which supports its mapping activities in Britain with revenues from private and public sources.[21] Another example is the National Aeronautics and Space Administration's partnership with Orbital Image Corp. on the Sea-viewing Wide Field-of-view Sensor satellite data.[22] License agreements are a natural way to implement such government–industry collaborations and can result in substantial cost reductions, albeit at the cost of some compromises in the data that are available to an agency.

Third, there may be insufficient funds in an agency's budget to support every worthwhile acquisition. In this case, the agency may decide that acquiring data through a license that provides very limited redistribution rights—despite its drawbacks—may be appropriate.

Fourth, although in theory every agency has an incentive to coordinate purchases with others in order to lower its costs, it may be prohibitively expensive to achieve the necessary coordination. Even where coordination is feasible, some agencies may be tempted to ignore potential savings because of institutional rivalries. In either case, the government may end up acquiring redundant or unnecessarily expensive data and some data may not be acquired at all. There may be cases where carefully crafted licenses, in which each of a number of agencies acquires limited rights in data, can overcome these difficulties.

6.4.2 Economic Disadvantages of Licensing

The principal economic disadvantage of obtaining data through licensing, rather than through outright purchase, is that the licensing restrictions limit the ability of an agency to distribute the data it has acquired in an economically efficient manner. Whereas efficiency in distribution

---

[20]At the limit, local governments would encourage very small groups to organize on their own.

[21]That is, an insurance company (Norwich Union), and the U.K. Environment Agency (testimony of Michael Bullock, Intermap).

[22]See Chapter 4, Section 4.2.1.

requires that no user who is willing to pay at least the incremental cost of distribution is excluded, that may not be possible when there are limits on the ability of any agency to redistribute data that it has acquired from the private sector. For this reason, licensing almost always sacrifices distributional efficiency by leading to overpricing (and undersupplying) geographic data. The amount of actual exclusion depends on individual circumstances. This effect is particularly important in cases where demand is price sensitive, that is, where many users are likely to drop out of the market each time prices move upward. Except for the rare case when every user is identical, or perfect price discrimination is possible, "deadweight loss" is unavoidable.

Government can almost always avoid the deadweight loss associated with exclusionary pricing and inefficiencies in distribution of information by using methods to procure data that permit unrestricted redistribution and providing these data to any user willing to pay the marginal cost of reproduction. Avoiding inefficiencies in production is likely to be more difficult. There is a broad consensus that governments should play a large role in the provision of geographic information, although the line between private and government provision remains contentious. The main challenge is to decide whether the advantages of licensing (efficiency in production) outweigh its drawbacks (efficiency in distribution). In the words of Thomas Jefferson, agencies should be careful to "draw…a line between the things which are worth to the public the embarrassment of an exclusive patent, and those which are not."[23]

6.5 NEGOTIATING THE LICENSE STRUCTURE: ECONOMIC CONSIDERATIONS

If an agency decides to acquire data through licensing, it must still negotiate terms. In this section, we note some economic considerations for the negotiations. The challenge facing agencies is to maximize the difference between benefits and costs. The nature of benefits will depend on the agency's goals, but would include the ability to use the data to fulfill the agency's mission at the lowest possible cost. The costs include royalties paid, the deadweight loss resulting from restrictions on government's ability to use and redistribute the data, and contract administration expense.

Like any private-sector business, agencies should insist on paying the lowest possible price for data, and should be willing to haggle or invoke

[23]Quoted in *Graham v. John Deere Co.,* 383 U.S. 1, 11 (1966).

competitive bidding to do so. Agencies should not attempt to ensure that private-sector firms receive sufficient royalties to "cover their investment" in data. Not only is such a policy unfair to taxpayers, but it sacrifices the supposed efficiency of using markets in the first place.

Agencies also must be willing to present creative solutions in which the "standard" package of license rights is adjusted to avoid paying for rights that the agency does not need. From a practical standpoint, agencies should seldom, if ever, pay more to license commercial data than to independently re-collect the same information.[24] Although this principle might seem to invite wasteful duplication, the threat will almost never be carried out. Instead, rational vendors will adjust their prices to stay competitive.[25]

Agencies can sometimes obtain reduced license fees in return from providing nonmonetary incentives, such as access to one-of-a-kind resources, to private licensors. Examples include access to raw and processed data in government archives, coordination with existing agency research programs, access to skilled agency employees, access to agency facilities, and possible synergies between private and public research agendas. In the New Economy, online "content" and the "ability to attract eyeballs" have also become assets. Agencies with prominent Web sites may be able to extract price concessions in return for posting advertisements or links to private vendors.[26]

6.6 LICENSING GOVERNMENT DATA TO THE PRIVATE SECTOR

We have already explained why efficient distribution requires government to make data available to all users who are willing to pay at least the marginal cost of distribution.[27] Nonetheless, there may be practical

[24]Agencies occasionally may pay a premium when there is no time to collect an independent dataset. With appropriate planning, such cases should be few and far between.

[25]More specifically, a rational vendor understands that "sunk costs are sunk." It is better to sell data at a loss than not to sell them at all.

[26]Advertising and links are a dominant source of revenue for private Web sites. In 2002, Yahoo earned roughly 63 percent of its first-quarter revenues ($192,700,000) from advertising (Subscriber Fees a Boost for Yahoo, *San Francisco Chronicle*, April 11, 2002, available at <http://sfgate.com/cgi-bin/article.cgi?f=/chronicle/a/2002/04/11/BU78117.DTL>.

[27]Agencies also should avoid restrictions on users' ability to create extensions and improvements to their data. This is true even when—as is usually the case—

obstacles to achieving this objective. First, transaction costs may make it impractical (i.e., uneconomical) to collect any fee, with the result that some users may obtain data that they value at less than the marginal cost of distribution.[28] Second, government databases sometimes include data that have been licensed from the private sector and the restrictions in those licenses prevent pricing at the marginal cost of distribution. Third, legislation or budget constraints may require some agencies to charge user fees that attempt to cover all or a portion of their data acquisition costs. Subject to these constraints, agencies should attempt to achieve efficiency in distribution by avoiding excessive user fees and unnecessary use restrictions.[29]

6.7 SUMMARY

Society makes geographic data investment decisions through two very different institutions: governments and markets. Deciding which sector should acquire and distribute a particular product has profound implications for economic efficiency. This chapter has reviewed the strengths and weaknesses associated with each institution. In general, markets are a good solution when the initial decision to invest in a particular product is controversial or uncertain. Conversely, government procurement is most useful when uncertainty about whether to invest is small, so that distributional efficiency becomes the dominant concern. Beyond these generalizations, additional considerations may apply to particular cases.

Agencies affect the government/market balance each time they acquire or distribute data. The challenge is to make these choices consciously with an eye toward economic efficiency. License design can be an important tool for setting this balance. For example, suppose that an agency

---

intellectual property rights protect the extensions and improvements [see footnote 7 in this chapter for related references]. Consumers who choose to pay for "improved" public data are better off than they would be in a world where intellectual property rights did not exist. Additionally, when government makes data widely available, the data products of would-be competitors are more likely to be competitively priced. The threat of actual or potential competition is a powerful constraint on the price that consumers pay for commercial data and services.

[28]This was the experience of the Maryland Department of Natural Resources (testimony of William Burgess).

[29]In addition, government agencies may be unable to engage in discriminatory pricing even when it would be efficient because legislation prevents them from doing so.

believes that certain geographic data products have uncertain value, so that investment decisions are best left to private markets, and that vendor pricing will not seriously limit society's use of whatever data are produced. Under these circumstances, the agency usually will wish to acquire data from private producers through licenses that give the agency modest use and redistribution rights. In general, such licenses promote markets by allowing the original suppliers to pursue additional sales due to the diminished likelihood that data provided to government will become widely available to others through actions of government.

Suppose, on the contrary, that the agency thinks that certain geographic data products have proven their worth, but that high prices are preventing many people from using them. In this case, the agency may wish to make the data it acquires widely available by acquiring them through licenses that give the agency broad redistribution rights.[30] Although such rights will limit any remaining private market for the product, that will be reflected in the price that the vendor demands, and the agency pays, for such a license. In return, efficiency in distribution is more likely to be achieved.

Finally, traditional licensing models are not the only—or, in some cases, the best—ways to promote economic efficiency. Other forms of private–public partnership are also possible. For example, some firms will not mount large data acquisition programs unless government agencies contribute resources. This can be done in a variety of ways, including cooperative research and development agreements, private–public partnerships, and licenses that obligate the agency to buy large volumes of data. From the agency perspective, such transactions may offer a good mix of efficiency in both production and distribution. Efficiency in production is achieved because the project must still realize significant commercial sales to be profitable. Efficiency in distribution occurs because the agency often has significant leverage to demand license terms that permit widespread dissemination on favorable terms, perhaps by requiring the vendor to donate its data to the public after a fixed period of years, or otherwise limiting the private partner's ability to impose high prices. Alternatively, government might bear the entire cost of data production and acquire unlimited rights in order to promote efficiency in distribution.

[30]In the limit, the agency may wish to purchase the data outright.

VIGNETTE F. A MAINSTREAM GEOGRAPHIC DATA MARKETPLACE DREAM

The geographic data marketplace, commonly referred to as Spatial Mart, has become a well-organized, convenient, and economical channel for reaching millions of customers. For many vendors, it is their primary distribution channel. Spatial Mart is a rich environment for commercial innovation and has gathered a critical mass of comprehensive content to meet broadening consumer demand. Innovators now focus on bringing new data products and services quickly and inexpensively to market rather than on maintaining complex selling and licensing arrangements.

All geographic data indexed by Spatial Mart adhere to licensing and metadata creation standards that ease automated search, retrieval, and financial functions. Additionally, Spatial Mart automatically tracks licensing terms in each transaction, and manages licensee payments and royalty distributions for any original and derivative works. Because of this, some commercial data suppliers now allow free downloads of substantial quantities of geographic data for product development purposes. These vendors' primary source of revenue is from commercial redistribution of derivative products. Thus, hoping that profitable uses of their data will arise, they allow free experimentation with their data and its incorporation in other products and services licensed through Spatial Mart.

Bob Nathan, who runs a trucking company, is searching Spatial Mart for all available geographic data meeting his technical and geographic coverage needs. After selecting a product, he clicks on a combination of standard licensing paragraphs to define his preferred data usage rights. These choices address such matters as the ability to use, alter, and disseminate the data, or to further create, license, and sell derivative products. Commercial vendors have already established prices that depend on various combinations of standard provisions and the volume and geographic coverage requested (data with few or no use restrictions typically are priced higher than those carrying substantial restrictions). Consequently, within seconds, Mr. Nathan receives a price quote and proceeds to the online purchasing step where he can comparison-shop—much like in the online purchase of airline tickets—if he chooses. He is satisfied with the quote in comparison with the other options, and acquires the data.

Had Mr. Nathan concluded that no offerings satisfied his desired combination of price and licensing terms, he would have been asked to offer the price, terms, and technical specifications he would accept. Any vendor interested in supplying existing data meeting Mr. Nathan's preferences could accept his offer—much like on priceline.com. Had Mr. Nathan needed custom data products or services rather than existing ones, Spatial

Mart's electronic bidding system would have recorded his technical and licensing preferences, any additional purchaser/vendor contract preferences, and then invited qualified vendors to submit their bids by a specified date.

Bob's dream is this: Can an operational infrastructure supporting efficient licensing and transaction interactions be developed that is open to all sellers and buyers of geographic data and services and will support an active and thriving marketplace?

7

The Public Interest

7.1 INTRODUCTION

The reasons for choosing a particular model for acquiring or distributing geographic data are not limited to legal rules and economic analysis. There are additional considerations that spring from collective values that fall outside legal mandates, regardless of economic efficiency. Some of these considerations were introduced in Chapter 2. They include, but are not limited to, fairness, fostering creativity, promoting democratic processes, personal security and freedom, and privacy.

This chapter discusses a sampling of public interest considerations that affect governmental policy for obtaining and sharing geographic data. Some factors suggest reasons to acquire geographic data outright, whereas others suggest that licensing data with use or dissemination restrictions is acceptable or even preferred. The discussion in this chapter is not intended to set forth an exhaustive list of issues that might be termed public interest.

7.2 PRESERVING AND ENHANCING THE PUBLIC DOMAIN

Historically, society has set limits on intellectual property to preserve a common "space" of ideas and information. The policy justifications for this public domain or "intellectual commons" include fostering public discourse, innovation, and equality.[1]

7.2.1 Public Discourse

Culture and politics depend on citizens' ability to obtain, display, and manipulate information. During the 1960s, for example, artists and activists used National Aeronautics and Space Administration (NASA) images of Earth to transform popular culture and advance the modern environmental movement.[2]

Government acquisition seldom reflects the entire value of publicly shared goods because the beneficiaries—members of the public and other tertiary users—do not sit at the bargaining table. Those who do—government agencies and vendors—can be expected to prioritize their own interests. In this environment, the benefits to the public generally may not seem worth the additional cost to an agency of acquiring full rights in data. This does not make the value of information for public discourse any less real or any less valuable. Basic geographic data and works may be essential to modern political and cultural debates.

7.2.2 Innovation[3]

Fostering creativity, whether scientific, artistic, or otherwise, requires the right balance between proprietary rights and free or open access to information. The ability of data providers to control and receive compensation has encouraged large numbers of vendors to enter the market. On the other hand, the availability at low cost of the U.S. Geological

---

[1]These goals are not entirely inconsistent with intellectual property and contract rights; the problem of striking the correct balance is covered in Chapter 6, Section 6.2. Our purpose here is to make explicit the benefits of a robust public domain.

[2]For example, the "earthrise" photographs from Apollo 8 gave a new global perspective of the planet (see <http://www.hq.nasa.gov/office/pao/History/Apollomon/apollo.html>).

[3]Chapter 6 discusses the role of public domain information in fostering further development of geographic data products (see in particular Section 6.2). Our purpose here is to make more explicit the benefits of public domain information, recognizing that the ultimate determination of when public domain information is desirable requires a balancing of costs and benefits.

Survey's (USGS's) topographic maps, without restrictions on reuse, has resulted in countless beneficial commercial and noncommercial uses.[4]

The benefits of a robust public domain for commercial and noncommercial innovation, like the benefits of informed public discourse, may be difficult to measure. Agencies also may lack sufficient incentives to fully account for them. Interest group politics[5] and agency capture,[6] although they should not be overrated, tend to divert attention from the concerns of unrepresented constituencies. The benefits of a robust public domain thus are likely to be undervalued, even though they are real and significant.[7]

Market forces are unlikely to capture the benefits of a robust public domain for basic research. Many research scientists, especially in universities and other nonprofit institutions, follow a fundamentally different mode based on open publication and shared discoveries.[8] Recent experiments with open source have extended this basic model to various nonprofit and commercial environments.[9] Although these developments are relatively new, they suggest that in some areas public domain innovations contribute significant benefits to society in both commercial and noncommercial contexts. Even if licensing and proprietary claims could accelerate commercial science, the negative impact on sharing within open communities still could reduce the total rate of discovery.[10]

---

[4]See National Research Council (NRC), 2002, *Research Opportunities in Geography at the U.S. Geological Survey*, Washington, D.C., National Academies Press, p. 62.

[5]See D.A. Farber and P. P. Frickey, 1987, The jurisprudence of public choice, *Texas Law Review* 65: 873.

[6]D. B. Spence and L. Gopalakrishnan, 2000, Bargaining theory and regulatory reform: The political logic of inefficient regulation, *Vanderbilt Law Review* 53: 599.

[7]As illustrated in a comparison of weather-related businesses in the United States and Europe—Chapter 4, footnote 95.

[8]Many institutions employing the same scientists assert intellectual property rights and licensing regimes to protect the fruits of basic research, however.

[9]See, for example, < http://www.opensource.org/> and < http://www.opengis.org/>. The open source community promotes open access to the program code that underlies software, free redistribution of the code, and allowance under some open source license arrangements of modifications of, and works derived from, the code.

[10]This problem is likely to be ameliorated where licensing regimes offer academic and nonprofit pricing that is at or near marginal costs, which is often the case. In addition to price discrimination, licensors have a range of licenses to reflect different user communities (e.g., Broward County, Florida, Property Appraiser's Office).

Discerning the appropriate balance is difficult because the impacts are not easily quantifiable. Even a small risk may be unacceptable.[11]

7.2.3 Equality and Equity

Government provides services and regulatory schemes that may give advantage to some citizens relative to others. In some instances, the rationale for such services and schemes is to correct inequities among citizens. Principles underpinning notions of equality and equity have some bearing on public geographic data. The purposes of these principles are several: they include creating equality of opportunity and providing access to information so that individuals can make informed decisions.

A shift to more licensing, with concomitant restrictions on data access and use, is likely to elicit more concern about inequality and inequity, particularly when differential pricing is employed. Strategies that may reduce claims of unequal treatment or unfairness[12] include designing decision-making procedures that make explicit the criteria on which decisions are based and that are nonarbitrary, and maintaining sensitivity to what stakeholders may perceive as rights to the status quo of freely available data. Historical agency practices regarding particular types of data, for example, might serve as a useful starting point for analyzing the significance of new data acquisition practices.

Conclusion: The federal government is one of the primary sources of and repositories for freely available or nominal-cost geographic data. Widespread acquisition of licensed data with use and dissemination restrictions opens the door to a shift in the balance of proprietary rights and free use and could have unforeseen and potentially harmful consequences. Careful consideration of data acquisition strategies can go far in preserving and enhancing the public domain in geographic data. When establishing data acquisition policies, agency mandates and missions may require agency policy makers to take into account the role of the federal government in maintaining and enhancing the public domain in geographic data.

---

[11]In the long run, a decreased public domain could depress commercial discovery. Many companies depend on universities for a steady supply of basic research.

[12]See E. Zajac, 2003, On fairness and self-serving biases in the privatization of environmental data, *in* NRC, 2003, *Fair Weather: Effective Partnerships in Weather and Climate Services*, Washington, D.C., National Academies Press, Appendix E at 209–212.

7.3 GOVERNMENT ACCOUNTABILITY AND TRANSPARENCY

Government accountability and transparency require agencies to ensure that the ability to control scarce geographic data never becomes "outcome determinative" for any political or judicial process. Minimizing use restrictions on geographic data facilitates citizens' ability to monitor government operations, a consideration that includes but is not limited to the legal requirements of the Freedom of Information Act (FOIA) and other legally mandated disclosure requirements.[13] Transparency is important to agency adjudications and rulemaking, to petitions to Congress for new legislation, and to mount court challenges to illegal government acts.

Society has an affirmative interest in demonstrating procedural fairness to every citizen. In some cases, this means putting basic data—census data, for example—into the public domain. In other cases, reasonable restrictions on access may be appropriate. For example, some states that use proprietary geographic software for redistricting limit access to the software and bundled data to a small number of physical locations.[14]

Licenses can be a useful tool for ensuring citizen access when geographic data are relevant to government action or policies. As a license condition, vendors might agree to keep products on the market to maintain citizen access, although this could become problematic when firms go out of business or produce multiple updates of the same product. In many cases, it may be more practical for government to acquire sufficient redistribution rights for individuals who seek access for political purposes, in which case rights can be narrowly defined in a license to contain costs. In other cases, it may make more sense to purchase unlimited rights.

Conclusion: Access to geographic data used by government is important for government accountability and transparency. Generally, agencies' acquisition of full rights will serve this goal. Licensing and more limited rights may suffice, as long as any restrictions on access to government data do not result in political or judicial outcomes that favor those with access over those precluded from access by the restrictions. When establishing data acquisition policies, agency mandates and missions may require that agency policy makers take into account the need for

---

[13]See Chapter 5, Section 5.4.2.

[14]Tim Storey, National Conference of State Legislatures, personal communication, January 2004.

accountability and transparency, obtaining full rights in geographic data that are needed for such purposes.

7.4 NATIONAL SECURITY

The goals of U.S. national security include enhancing American military capabilities and diplomatic influence while denying similar capabilities to hostile governments and terrorists. Among the options for satisfying these goals are preserving access to essential remote-sensing assets, denying satellite data to potential adversaries, and restricting access to other geographic data.[15] National security priorities change over time, and data declassification also has potential benefits.[16]

7.4.1 Preserving Access to Essential Assets[17]

National militaries have always drawn strength from their respective economies. This interdependence seldom raises national security concerns as long as goods and services are available from multiple sources, or can be duplicated in a reasonable length of time. However, some civilian assets are so expensive and complex that governments have no hope of replacing them in an emergency. Governments traditionally have reserved the right to commandeer such assets.[18] Prior to the mid-1990s, however, few, if any geographic data collection assets fit this description.

The rise of commercial satellite remote sensing over the past decade has blurred the line between civilian and military geospatial assets. The typical imaging satellite costs well over $100 million and takes up to five years to design, fund, build, and launch. This means that the nation's satellite imaging capabilities are essentially fixed in the short run. Additionally, the Department of Defense (DoD) has become more dependent on com-

[15]We also recognize that not all strategies for achieving these goals are acceptable. We examine privacy constraints later in this chapter.

[16]Approximately 880,000 photographic images that were taken between 1959 and 1972, primarily collected by the CORONA satellite series, have been declassified and are publicly available. See <http://edc.usgs.gov/products/satellite/declass1.html>.

[17]Arguments for policies that ensure the availability of geographic data for national security could apply as well to data needed for other purposes.

[18]Recent examples include the U.S. military's use of commercial airliners to ferry troops to the Persian Gulf and Britain's use of the *QE2* as a troopship during the Falklands conflict.

mercial images to supplement and extend its own in-house capabilities.[19] This trend is likely to accelerate. The military's demand for satellite images is also likely to grow during wartime and national emergencies, and commercial satellite imagery is likely to become more important in meeting homeland security needs, especially at the state and local level.[20] Guaranteed access to commercial assets is important for obtaining the needed capability. The government can no longer afford to let civilian assets "go dark."

Some military requirements are met by statute. For example, all U.S. commercial satellite operators must receive government operating licenses[21] and commit to various national security goals. Although operating licenses have been an important mechanism for achieving national security goals, the system is inherently limited because there are practical limits to what the government can demand; if authorities impose too many requirements, firms will go out of business or leave the industry.

Data acquisition agreements—including, but not limited to, licenses—add flexibility that may not be possible through operating licenses. They provide formal, contractual guarantees that government will be able to purchase images. The prospect of future acquisitions provides a "carrot" that military planners can trade on to extract special treatment during an emergency.[22] Furthermore, such agreements provide a vehicle for subsidizing an industry that might not survive on its own. Although the practice cannot be justified on purely economic grounds,[23] subsidies serve national

---

[19]DoD has discussed various strategies for "piggybacking" its needs on the commercial sector. These include increased purchasing of commercial images and leasing commercial satellites so that they can be redirected to military objectives. J. Singer, 2002, Changes ordered in classified security program, *Space News* (Nov. 25), available at <http://www.space.com/spacenews/>; J. Singer, 2002, NRO faces potential gap in satellite coverage, *Space News* (July 15), available at <http://www.space.com/spacenews/>.

[20]See S. D. Kuo, 2003, Homeland issues in the use of space assets for homeland security, *Air and Space Power Journal* (Spring), available at <http://www.airpower.maxwell.af.mil/airchronicles/apj/apj03/spr03/spr03.html>.

[21]Not to be confused with the licenses that are the subject of this study.

[22]Such as the option to "commandeer" a satellite.

[23]Most economists reject so-called "infant industry" arguments that favor government protection or subsidies for private ventures. Such arguments uniformly assume that government can "pick winners," that is, make investments that consistently outperform private venture capital markets.

security interests by mitigating some of the risk that DoD's Future Imagery Architecture program may not succeed.[24]

7.4.2 Denying Satellite Data to Potential Adversaries

The ability to restrict access to commercial satellite imagery becomes particularly valuable during wartime and national emergencies. Restricting access using contractual vehicles has been preferred over implementing regulatory control as allowed in the satellite operating licenses. In the long run, such formal restrictions may be less important—and certainly less flexible—than judicious exercise of government purchasing power. For the foreseeable future, U.S. commercial satellite companies need massive federal government acquisitions to survive.[25] This provides a powerful incentive to agree to withhold images from hostile powers. During the war in Afghanistan, Space Imaging, Inc., entered into a contract with DoD that assured DoD of access to all images from the war zone, thus restricting access by adversaries (and the Media) to intelligence that could have been used to track U.S. forces.

Conclusion: Government geographic data acquisition practices play an important role in ensuring that data products supporting national security, particularly satellite imagery, remain available to U.S. intelligence agencies and the military, but unavailable to adversaries. When establishing data acquisition policies, agency mandates and missions may require agency policy makers to take into account the need to ensure that geographic data resources needed for national security remain available to the government and unavailable to adversaries.

7.4.3 Restricting Other Geographic Data

When restricting access to geographic data, governments must balance legitimate uses against the possibility of abuse.[26] The former Soviet

---

[24]J. Singer, 2002, NRO faces potential gap in satellite coverage, *Space News* (July 15), available at <http://www.space.com/spacenews/>. See also discussion of the 2003 Commercial Remote Sensing Policy in Section 5.4.1.4 of Chapter 5.

[25]Clearview and Nextview are two recent examples of National Geospatial-intelligence Agency contracts with industry (see Appendix D, Section D.3).

[26]See J. C. Baker, B.E. Lachman, D.R. Frelinger, K.M. O'Connell, A.C. Hou, M.S. Tseng, D. Orletsky, and C. Yost , 2004, *Mapping the Risks: Assessing*

Union routinely suppressed geographic data showing roads, factories, and entire towns. By contrast, U.S. military strength rests on a market economy, decentralized decision making, and individual initiative. In these circumstances, it is counterproductive to restrict geographic data dissemination unless the potential harm substantially outweighs legitimate uses. Additionally, classification conflicts with the public's "right to know." This leads to a presumption against classification in ambiguous cases.

In practice, few publicly available geographic data products raise national security concerns. For example, 6 percent of 629 federal datasets were judged by Baker et al. (2004)[27] to be potentially useful to attackers and 1 percent was both useful and unique. Nonetheless, the events of September 11, 2001, placed new emphasis on low-technology threats. Today, sensitive geographic information includes building designs, secure locations, utilities and power lines, hazardous materials facilities, and other pieces of "critical infrastructure."[28] Many of these facilities are privately owned, and government often must share information with or acquire information from private entities. Consequently, agencies need mechanisms for controlling distribution to large (but select) groups of individuals and entities within and outside the government. The principal reaction has been to accelerate the development of a so-called "sensitive but unclassified" category for government data.[29] In 2004, the Department of Homeland Security initiated a "Protected Critical Infrastructure Information Program"[30] that solicits potentially sensitive information from private and other sources and, if the information qualifies for "protection," restricts its distribution through such means as providing immunity from FOIA requests.

---

the Homeland Security Implications of Publicly Available Geospatial Information, Santa Monica, CA, Rand Corp.

[27]See footnote 26.

[28]See Executive Order 13010 on Critical Infrastructure Protection, 61 FR 37347, July 17, 1996.

[29]Congressional Research Service, 2003, "'Sensitive but Unclassified' and other federal security controls on scientific and technical information: History and current controversy," available at <http://www.ieeeusa.org/forum/REPORTS/RL31845.pdf>. Information is classified on the basis of well-established criteria and procedures. "Sensitive" information, on the other hand, is information that does not meet the criteria for classification but that arguably might be useful to terrorists. It is a discretionary category, the application of which has been broadened since the September 11, 2001, terrorist attacks. *Id.*

[30]See < http://www.dhs.gov/dhspublic/display?theme=92>.

Licensing restrictions provide a natural mechanism for controlling data after they leave government. In principle, agencies could use licenses to delegate authority. They could, for example, let private entities decide how to redistribute sensitive data affecting their installations, although this strategy might let private entities restrict access for improper reasons, such as avoiding public scrutiny. Licenses also could be used to allow academic researchers, journalists, and other prescreened individuals to have access.[31]

Conclusion: When deciding whether to restrict geographic data access on national security grounds, policy makers must carefully weigh the need for restricted access against the public's interest in being informed and having access to important information about their communities and environment. When restrictions are appropriate, licenses can be vehicles for structuring data access.

7.4.4 Declassification Policy

Since the 1960s, the U.S. military has invested enormous resources in imaging, acoustic, seismographic, and electromagnetic surveillance of Earth's land surface and oceans. The resulting datasets have substantial value for academic research and certain political issues (e.g., global climate change).[32] Declassification has accelerated since 1995 when

---

[31]See Congressional Research Service, 2003, at p. 43.

[32]See, for example, NRC, 2001, *Resolving Conflicts Arising from the Privatization of Environmental Data*, Washington, D.C., National Academies Press, p. 66. This report describes the societal value of such declassified data as (1) geodetic data that the civilian community can now use in ocean floor studies, (2) submarine data on Arctic Ocean ice-cover thickness changes over time, and (3) high-resolution Cold War spy satellite data from CORONA satellites. For a discussion of the value of declassified data in land surface change studies, see NRC, 2002, *Down to Earth: Geographic Information for Sustainable Development in Africa*, Washington, D.C., National Academies Press, p. 87. Declassified geographic data, like any recovered geographic data, are only useful if the accompanying supporting information (e.g., metadata) provides sufficient context and the data are not disrupted (e.g., by cloud cover). As of June, 2004, roughly 2 percent of declassified images in the "Declass 1" holdings of EROS Data Center were missing coordinates and roughly 40 to 50 percent of the images contained significant amounts of clouds obscuring the view of the land surface or had inherent contrast problems (John Faundeen, USGS, personal communication, June 2004).

President Clinton signed two related Executive Orders.[33] Potentially, such data also could boost the commercial satellite industry by demonstrating the value of satellite images to possible customers, and providing a high-resolution baseline against which current images can be compared to detect changes. We have argued that the existence of significant national security concerns can override the economic efficiency arguments that favor widespread distribution of government data. The reverse is also true: Once national security concerns fade, government must act decisively to resume widespread distribution.

Conclusion: When establishing data acquisition policies, agency mandates and missions may require agency policy makers to take into account the need to declassify classified geographic data to make them widely available when protection is no longer warranted.

7.5 FOREIGN POLICY

Geographic data have important value in establishing and maintaining foreign good will, improving trade, enhancing intergovernmental relationships, and assisting in the democratization of foreign institutions through making data that have economic, political, and environmental value available to other nations.[34] For example, the foreign policy value of remote-sensing data was recognized as "a very sensitive matter involving agreements which have been reached over the years between the State Department and other nations."[35] In some important cases, geographic data and services are subject to U.S. obligations under international law and nondefense foreign policy obligations.[36]

---

[33]See <http://www.fas.org/sgp/clinton/eo12958.html>, <http://www.fas.org/irp/offdocs/eo12951.htm>. See also NRC, 2001, *Resolving Conflicts Arising from the Privatization of Environmental Data*, Washington, D.C., National Academies Press, p. 66.

[34]See also the 2003 Commercial Remote Sensing Policy, discussed Chapter 5, Section 5.4.1.4.

[35]See *The Land Remote Sensing Policy Act of 1991*, 102d Congress, 2d Session, at 210 (1992).

[36]See, for example, Canadian Space Agency, 1994, *Radarsat Data Policy*, RCA—PR0004, § 10.1.b., at p. 11.

Conclusion: Geographic data access may serve important foreign policy goals that should be taken into account by agency policy makers in establishing frameworks for data acquisition agreements.

7.6 LAW ENFORCEMENT

Law enforcement uses geographic data and GIS in applications such as tracking patrol cars, locating "911" callers, searching for illegal drug production, and profiling the likely locations of snipers. As with national security, most of these law enforcement applications take advantage of information goods developed for civilian markets. Unlike national security, law enforcement seldom needs access to unique and expensive assets such as satellites, and can usually be thought of as one of many end users.

Law enforcement has an interest in restricting access to "sensitive" geographic information. In at least one instance, burglars have tried to access tax assessor files to find high-value homes that lack security systems, and poachers have used environmental data to locate rare and endangered species.[37] As with national security, government must weigh potential dangers against legitimate use. The public's right to know implies a presumption against restrictions, absent clear evidence of need.

When restrictions are needed, blanket restrictions should be the exception rather than the rule. In many cases, modest restrictions on anonymity and convenience are sufficient; for example, would-be users can be required to present identification or undergo a formal background check. When geographic data are made available outside government agencies, licenses can support such procedures by imposing redistribution restrictions that range from "no distribution without express government permission" to "distribution at recipient's discretion." The latter option may be particularly useful when government relies on the recipient to help protect privately owned assets.

On the other hand, it is also important for law enforcement agencies to keep in mind that government use of licensed data and services must be consistent with Fourth Amendment protections against unreasonable search and seizure and that license terms cannot be used to circumvent them.[38] Depending on the license terms and the relationship between the

---

[37]Testimony of Randy Johnson, Hennepin County, Minnesota.

[38]We do not attempt to discuss in any depth the legal constraints on obtaining and using information gained from aerial photographs and satellite imagery. Such methods may implicate Fourth Amendment or other concerns. In *Dow Chemical Co. v. United States,* 106 S. Ct. 1819 (1986), the Supreme Court

government and the private entity, the private entity's activities might constitute state action, subjecting it to unintended liabilities. Contracting parties should consider these facets of the Fourth Amendment when law enforcement is involved and seek counsel when a license is being negotiated.

Conclusion: Law enforcement agencies have an interest in restricting access to data that may be of interest to potential lawbreakers. Nonetheless, the public's right to know implies a presumption against restrictions—particularly blanket restrictions—absent clear evidence of need. And law enforcement agencies should not impose restrictions on public access to geographic data except when there is clear evidence of such need.

7.7 PRIVACY

Aerial photography and satellite imagery reach into backyards and businesses.[39] Government agencies, commercial firms, and individuals routinely use these data to conduct surveillance. When government agencies or their agents, which may include contractors, engage in such activities, it may implicate the Fourth Amendment, as discussed earlier, or it may implicate privacy concerns.[40] Private parties who engage in such surveillance may also incur tort liability for invasion of privacy.[41]

---

held that aerial photography of Dow's chemical plant for an environmental inspection did not constitute a search for the purposes of the Fourth Amendment and that no search warrant was required. The Fifth Circuit took a different view of the use of aerial photography in industrial espionage in *E. I. duPont deNemours & Co. v. Christopher,* 431 F.2d 1012 (5th Cir. 1970), *cert. denied*, 400 U.S. 1024 (1971), holding that obtaining information on trade secret chemical processes through aerial photography of a plant under construction was improper.

[39]For a discussion of improved sensing capabilities and privacy concerns, see E. T. Slonecker, D. M. Shaw, and T. M. Lillesand, 1998, Emerging legal and ethical issues in advanced remote sensing technology, *Photogrammetric Engineering and Remote Sensing*, 64(6): 589–595.

[40]Many federal statutes, for example, provide for protection of citizens' privacy. See, for example, 12 U.S.C. § 2803 (requiring deletion of personal identifiers from public repositories of home mortgage information).

[41]Actress Barbara Streisand sued a photographer who posted an aerial photograph of her Malibu, California, home on a Web site along with 12,000 other pictures of the California coast, claiming invasion of privacy. The court rejected her claim, however, indicating that the California coast is a matter of

Marketing firms use geographic attribute data to infer religion, ethnicity, buying habits, and political preferences. Using geographic data for such purposes entails both costs *and* benefits. For example, marketing data intrude on privacy by revealing attributes that consumers might prefer to keep anonymous, but also deliver goods and services to people who want them. The balance is not obvious and differs from person to person.

In principle, consumers should strike this balance for themselves. In practice, consumers rarely have enough time, energy, or information to give meaningful consent. Government fills this vacuum by imposing limits based on assumptions about what a *reasonable* consumer *would have* asked for. In the case of geographic data, government may protect privacy by publishing data with degraded spatial resolution that obscures individual households. Examples include census returns, farm statistics, and soil surveys. The system is not perfect, however. Commercial and academic users have become adept at teasing individualized data from government figures.[42] This activity represents the worst of all worlds: breached privacy *and* a social investment in bypassing government safeguards.

Degraded resolution imagery and aggregated statistical data may be the only viable options when *any* use of household-level data would violate privacy. However, some uses may not violate privacy. For example, academic researchers often seek to establish statistical regularities from household-level data in which the identity of individual households remains anonymous.[43] In this case, licenses granting researchers access to undegraded data in return for contractually defined limits on reuse and redistribution could suffice. In some cases, formal license restrictions may provide a more effective barrier against misuse than technical protections based on blurred or aggregated data.

Conclusion: When establishing data acquisition policies, agency mandates and missions may require agency policy makers to take into account the need to ensure the protection of privacy.

---

public interest and that the photography was not sufficiently intrusive to be actionable (K. R. Weiss, 2003, Judge rejects Streisand privacy suit, *Los Angeles Times* [Dec. 4], p. B1).

[42]M. Monmonier, 2002, *Spying with Maps: Surveillance Technologies and the Future of Privacy*, Chicago, University of Chicago Press.

[43]Researchers sometimes need the householder's identity to match government information with other databases. The resulting breach of privacy may or may not be acceptable, depending on the facts of each case.

7.8 PROMOTING WIDESPREAD USE OF GOVERNMENT DATA

Effective access to readily found archived data supports the public interest in many ways, including reducing redundant data collection at taxpayer expense, enabling unanticipated uses of data, and allowing detection of changes on Earth's surface in support of land management, global environmental change research, and other applications. There are several "infrastructure" systems and functions underlying these capabilities. These include systems that ensure long-term preservation and access to data (including licensed data) and functions relating to standards setting, data discovery, and data sharing. Licensing terms to or from government can influence whether and to what extent government data are reused.

7.8.1 Finding and Sharing Data

Data that cannot be located might just as well not exist. For this reason, spatial data infrastructures (1) encourage the use of "metadata" that provide a standardized, shorthand description for each geographic data file,[44] and (2) strive to ensure that there are sufficient venues such as catalogs, clearinghouses, Internet portals, and institutions such as libraries where would-be users can search for and acquire existing data.

Because of the public interest in geographic data use, government has a responsibility to promote metadata and clearinghouse activities in the broader society. To some extent, government cannot help influencing the rest of society; government data purchases and licensing agreements are often so large that private and nonprofit providers have a built-in incentive to make their own institutions and standards conform. In other cases, government may consciously intervene to foster communitywide strategies for locating and exchanging data.[45]

Although government agencies have primary responsibility for government data, they also rely on commercial firms to perform metadata or

---

[44]Metadata is information about data; for example, it might record such details as the collector, the sensor used, and when the data were collected (see Federal Geographic Data Committee, 1998, *Data Content Standard for Digital Geospatial Metadata*, available at <http://www.fgdc.gov/standards/documents/standards/metadata/v2_0698.pdf>.

[45]For example, the Federal Geographic Data Committee's work on metadata standards, or the Geospatial One-Stop initiative to develop an Internet portal (see < http://www.geodata.gov/>).

clearinghouse functions that facilitate sharing among agencies. To date, this strategy has been mostly ad hoc. Agencies should consider making metadata and the disposition of the data an explicit part of future licenses when it is cost-effective to do so.

7.8.2 Standards

Compatible standards increase the likelihood that geographic data will be effectively reused. Since no single standard is best for all purposes, some fragmentation is almost always efficient. Nonetheless, network effects imply that most new standards must attract a "critical mass" of users before they are useful. In these circumstances, change may require a coordinated, near-simultaneous migration by large numbers of users. Strong leadership and institutions can facilitate this process.

Some government intervention is unavoidable, and government decisions can and do influence users throughout the economy. Government may also *choose* to intervene by leading communitywide efforts to design and adopt new standards. Such initiatives may be necessary where cultural, political, and network effects have blocked adoption of technically superior standards. In such cases, agencies should foster consensus, not create it.

7.8.3 Archives

Agencies frequently must decide whether and how to archive data. In some cases, long-term data access supports the agency mission. Most government archives house data originally collected by or for government,[46] but commercial satellite licenses blur this distinction of government-only data by requiring vendors to inform agencies before privately owned data are deleted—in which case government may take over data archiving responsibilities to support ongoing missions or the public interest.[47] For example, USGS's Sioux Falls archive houses satellite data collected by the French company Systeme Probatoire Pour l'Observation

---

[46]For example, NASA's Distributed Active Archive Centers archive data from NASA and partner agency missions. For details of their holdings, see, NRC, 1998, *Review of NASA's Distributed Active Archive Centers*, Washington, D.C., National Academies Press.

[47]See <http://edc.usgs.gov/archive/ceos//data_purge_alert.html> for a list of datasets that USGS's EROS Data Center is aware of potentially being "purged."

de la Terre.[48] In principle, there is no reason why government archiving should be limited to data that agencies have previously owned or licensed. And government may intervene as the "archiver of last resort" for data that otherwise would be lost—as is the case with some data at the National Satellite Land Remote Sensing Data Archive.[49]

Conclusion: Government serves a critical public interest role by encouraging geographic data reuse through archiving and sharing, facilitating consensus formation on standards, and working with commercial vendors to preserve potentially valuable data. Licenses that enable government to support this role serve the public interest.

Conclusion: In principle, government can promote reuse of geographic data by negotiating licenses that limit commercial firms' ability to discard data prematurely,[50] promote uniform and high-quality metadata, and encourage standards that make geographic data interoperable with a wide range of hardware, software, and data products. Alternatively, government can promote widespread data reuse through actions that support marketplace innovation. By acquiring full ownership rights in the geographic data it acquires, government can provide unfettered access to all users. The ability of anyone to fully scrutinize and freely experiment with the data, along with the lack of need to pay royalties, enhances its reuse as raw material for value-added activities by commercial firms, government agencies, and academic researchers.

Conclusion: When establishing data acquisition policies, agency mandates and missions may require policy makers to take into account the need to promote data sharing and reuse, development of consensus standards, and archiving of data.

7.9 SUMMARY

Public discourse, equality, and innovation are benefits that are not easily assessed but accrue to society as a whole. These benefits have been well served by public domain data, which have been the norm

---

[48]See <http://edc.usgs.gov/archive/nslrsda/>.

[49]*See* <http://landsat7.usgs.gov/datatrans.php> for archiving policy for Landsat 7.

[50]This is currently achieved through operating licenses, but government could, in principle, negotiate similar provisions in data licenses.

under a legal regime in which geographic data, once published, were free for anyone to use. Such data also serve government accountability and transparency, although some license restrictions may also support these public interests in some cases. National security, foreign policy, law enforcement, and privacy issues present common challenges to policy makers considering geographic data access issues: how to weigh potentially harmful or intrusive uses against legitimate uses. Blanket restricttions and classification on national security or law enforcement grounds are inadvisable except in unambiguous cases. Furthermore, because of the potential benefits of classified data beyond the national security arena, timely declassification is important. When classification is necessary, licenses can be used to limit access to specified users. Government also can use licenses to promote reuse of geographic data by negotiating terms that limit commercial firms' ability to discard data prematurely, promoting uniform and high-quality metadata, and encouraging standards that make geographic data interoperable across a wide range of hardware, software, and data products.

In the next chapter, we integrate public interest, legal, and economic considerations into a process for deciding when licensing to or from government may be appropriate. We also discuss how license-based approaches may best serve stakeholders.

VIGNETTE G.
A GLOBAL INFORMATION COMMONS DREAM

Betty is a town engineer. For the past 15 years she has compiled detailed maps and affiliated digital records on the locations, sizes, materials, and conditions of all storm and sanitary sewers, waterlines, powerlines, waterways, buildings, and streets in her community. Jack is a local college professor. For the past 30 years his hobby has been to visit urban and rural environments throughout his state to find lichens that he then precisely locates, identifies, and documents. Betty and Jack meet in the local coffee shop and discover that they have much in common.

Betty and Jack are regularly asked for copies of the geographic and affiliated datasets they have compiled. Both remark that they are tired of responding to requests and would be more than willing to make their files widely available but only under certain conditions. Both would like to be able to reliably retain credit and recognition for their contributions to the public commons. It is fine if others use their work but they would like to be acknowledged. Further, for their data files to be useful to others, metadata needs to be provided. However, creating meaningful metadata for geographic data files is burdensome and needs to be made much easier. Since the town's datasets are often used for making decisions, Betty would like to see a means for limiting the town's liability for any files that are made widely and openly available on the Web. Both like the prospect of many others benefiting from their data files. They are also attracted by the prospect of others reviewing their work and suggesting additions or improvements. Both believe that their files might be of value many years into the future, but whether their works will still be available is an open question.

While they act locally, Betty and Jack contemplate a "global commons" consisting of a broad and continually growing set of freely usable and accessible geographic datasets. They believe the dream could be made real by providing the ability for geographic data contributors to quickly and easily create metadata and open access licenses. The system would allow delivery of files to a permanent online environment where the files could be readily found and retrieved. The basic purpose of the license would be to provide an easy, affirmative legal mechanism by which they could make known to the world that their data files are available for use without the law assuming that the user must first acquire permission. The license would ensure that the originator and all value-adders have a legally enforceable right to credit for their work, liability exposure would be substantially reduced through the license provisions, and the license could prevent, if desired, the efforts of the originator and value-adders from being captured as the intellectual property of others. Perhaps the system might also embed technological identifiers or protections in the files.

As they get up to leave, Jack remarks that if such a system could be developed, there are probably tens of thousands of individuals similar to themselves already creating valuable geographic data who would now have sufficient means and incentives to make their files available to the rest of the world. Betty agrees.

8

Licensing Decisions and Strategies

8.1 INTRODUCTION

Drawing on the analyses from the preceding three chapters, this chapter presents a menu of options that government may use to decide whether and how to license geographic data and services from and to the private sector. For cases in which licensing is appropriate, the chapter addresses ways to make licensing a more versatile tool, and explores license-based approaches that might satisfy the broadest range of stakeholders.[1] There is no intent by the committee through the presentation of this material to convey explicitly or implicitly a general preference for or against acquisition of geographic data and services through licensing.

The chapter begins by presenting a framework for, and issues to consider in, determining whether licensing may be appropriate in specific instances. We then propose strategies that agencies can use when acquiring or distributing data under license. These strategies can be implemented immediately (we discuss longer term strategies in Chapter 9). We focus primarily on the case in which government licenses data from private vendors. However, Section 8.4.3 examines the opposite case of the agency as licensor.

---

[1]Item 4 of the committee's task. Chapter 9 also addresses aspects of this task.

8.2 A FRAMEWORK FOR AGENCY DECISION MAKING

When acquiring or distributing geographic data, agencies must make choices about the most effective and efficient way to accomplish their missions, weighing such factors as data cost, quality, and fitness for use. In making these choices, agencies need to be clear about (1) what the data are initially needed for and (2) what follow-on applications are required or desirable, including both discretionary and nondiscretionary applications. A number of considerations affect the decision-making context for agencies. These can be conveniently presented as a series of iterative steps (Figure 8-1).

1. **Delivering on an agency mission**
    a. required (by law) and desired (by choice),
    b. accommodating fiscal realities, and
    c. balancing competing interests internal to government,

    in a context that

2. **strengthens government accountability**
    a. respecting the public right to know and understand decisions,
    b. supporting efficiency and effectiveness of government operations, and
    c. promoting flexibility of government choices (past, present, and future),

3. **licensing represents a different and potentially valuable tool**

    with the understanding that

4. **licenses have advantages and disadvantages**

    taking into account that

5. **public interest must be part of the decision-making, and**

6. **government choices affect "the marketplace."**

FIGURE 8-1 Steps in the decision-making process for geographic data acquisition and distribution. Arrows indicate points in the decision process when it is valuable to revisit earlier components of the decision sequence.

As indicated in Figure 8-1, licensing represents a new and potentially valuable tool for accomplishing agency missions. Like any powerful tool, licenses have both advantages and disadvantages. Before deciding for or against licensing, government decision makers should have a clear understanding of how a decision to license is likely to affect the goals of efficient and accountable government, the broader public interest, and the probable impact on private markets.

8.3 GOVERNMENT'S MISSION: CLARIFYING AND ACHIEVING THE GOALS

Like most contracts, the structure of a license is a matter of negotiation between license holders and licensees.

Recommendation: Before entering into data acquisition negotiations, agencies should confirm the extent of data redistribution required by their mandates and missions, government information policies, needs across government, and the public interest.

In this section, we suggest rules of thumb that agencies can use to refine geographic data acquisition decisions in support of their mandates and missions. We begin by describing how government could decide which data to acquire and conclude by asking when acquired data should be redistributed.

8.3.1 The Procurement Mission: When Should Agencies Choose Licensing?

Traditionally, the federal government's preferred procurement method has been to acquire geographic data unencumbered by restrictions on use or reuse.[2] However, depending on the circumstances, the advantages of licensing may outweigh the social and economic drawbacks of acquiring restricted geographic data. Because beneficial downstream uses and the public's interest in the free flow of information cannot be fully anticipated, agencies should exercise caution in construing their mandates and missions to permit licenses that restrict such uses.

---

[2]Because federal agencies do not have a proprietary interest (i.e., copyright) in government data and records, there has been little past need within government to bear the economic burden to support operations protecting proprietary interests. Additionally, licensing restrictions were not prevalent in the private sector until the early 1990s.

Agencies need data to support a wide variety of mandates and missions.[3] For this reason, there is no simple answer to the question of when agencies should choose to acquire geographic data through restrictive licenses. Licenses can be made to work in most cases in which geographic data are needed; the challenge lies in determining when licensing data subject to use restrictions is superior to in-house production, purchase of unencumbered data, or acquisition through grants, prizes, cooperative research and development agreements (CRADAs), partnerships, or other alternative methods.

One logical starting point is to identify the method that achieves agency missions at the greatest excess of benefits over costs.[4] Licensing may or may not be this method. The following four examples illustrate some advantages and disadvantages of licensing.

1. *Time-Sensitive First Response.* As discussed in Chapter 4, Section 4.2.5.1, licensing data from the private sector can sometimes result in substantial benefits at minimal direct cost. For example, Palm Beach County, Florida, and the Federal Emergency Management Agency (FEMA) both used licensing to acquire data more quickly than they otherwise could have.

2. *Transaction Costs of Licensing Relative to the Value of the Data.* Licensing may have significant limitations in some circumstances, particularly for small-scale projects in which transaction costs for negotiating and administering contracts may be prohibitively expensive in light of the size of the overall project.[5] In these

---

[3]As used in this discussion, a *mandate* is a required function that is defined by statute, administrative code, or case law. A *mission*, on the other hand, is either a discretionary function or is an approach to accomplishing a mandated function that is carried out as part of a strategic or operational direction. A strategic direction must necessarily accommodate the overall information policies of government as expressed in its laws and regulations.

[4]This is only a starting point since agencies must address more than their own parochial interests; they need to consider the full range of interests encompassed by the legislated mandates and missions. A comprehensive benefit-cost analysis would include both short-term and long-term perspectives and consider both quantifiable and nonquantifiable factors.

[5]This applies when licenses must be individually negotiated. It would not apply to click-wrap or other forms of mass licensing. Transaction costs for licensing tend to be extremely varied—sometimes greater, and sometimes less than for other procurement methods—in part because of the geographic data community's relative inexperience in working with licensing environments compared with its experience with other methods.

circumstances, outright data acquisition is often preferable to license-based solutions since the need to negotiate and then administer the license over the life of the data may create high transaction costs.

3. *Regulatory Data.* In cases in which public data are used to develop and implement regulations, affected parties and the general public have a strong interest in gaining access to whatever data are needed to validate and/or appeal agency actions. Licensing data from the private sector subject to restrictions on downstream access or use may have limited value in this regard.

4. *Operational Data.* Governments may use data to support such functions as dispatching maintenance staff or inspectors, automatically tracking vehicle location, or managing mobile assets. Such functions may have little policy or regulatory impact. Consequently, government acquisition of commercially licensed data that allows the public and news agencies to view but not acquire data may be a viable alternative. However, citizens may be interested in learning or need to know the details of a specific operational situation (e.g., the dispatching events on September 11, 2001) in order to understand how the policies and operations of government affect its ability to perform.

Recommendation: Agencies should experiment with a wide variety of data procurement methods in order to maximize the excess of benefits over costs.

8.3.2 The Dissemination Mission: When Should Government Acquire Unrestricted Data?

This section provides guidance to agencies deciding when to acquire geographic data with no restrictions on reuse and distribution.

The first consideration is whether the data have regulatory or policy consequences and should therefore be available for redistribution subject to few or no restrictions (Section 8.3.2.1). If this is *not* the case, then the decision depends on whether the data are for internal use only (Section 8.3.2.2), are to be widely distributed at marginal cost (Section 8.3.3.3), or are being acquired for limited redistribution (Section 8.3.3.4). Most opportunities for creative licensing solutions lie in this last-mentioned category of limited redistribution.

The main determinant of whether an agency should acquire geographic data subject to restrictions on reuse or redistribution is the agency's interpretation of its mandates and missions. Factors that agencies should take into account when exercising discretion are discussed later.

8.3.2.1 Data of Regulatory or Policy Importance

When government uses geographic data to promulgate regulations, formulate policy, or take other actions that affect the rights and obligations of citizens, there is a compelling interest in making these data available so that the public may understand, support, or challenge government decisions.[6] This interest often will be served by acquiring unlimited rights in data, but also may be accommodated in some circumstances by licensing data under conditions that permit access for more limited purposes. For example, in some cases, the public may need access only to views derived from satellite data, rather than the underlying data themselves. The important principle is that access to information cannot be so limited, its distribution so difficult, or its content so closely held by government that outcomes of political debates are determined by unequal access to data.

8.3.2.2 Internal Use

When geographic data are used for purely operational tasks (e.g., emergency dispatch, project management–style resource allocation, traffic management, military operations), distribution may not be necessary. When such circumstances do arise, the agency is like a private business—it should adopt licensing solutions if they are more cost-effective than the alternatives.

---

[6]The Freedom of Information Act (FOIA) cannot be used to compel disclosure of legitimate trade secrets or proprietary information. However, an agency may not be able to support its decisions in court or elsewhere if the public or a court cannot scrutinize the basis of an action.

8.3.2.3 Economic Factors Supporting Broad Distribution of Government Data[7]

There are circumstances when providing access to the public at no more than the marginal cost of distribution is appropriate from an economic perspective and others when it is not. Before deciding to accept licensed data subject to some level of reuse restrictions, agencies will need to assess the potential value of the data to secondary and tertiary users against the additional costs of obtaining unrestricted rights. Because the benefits of secondary and tertiary uses may be difficult to quantify, agencies should guard against the temptation to accede to reuse restrictions too readily. Conversely, there is also a danger that government may avoid licensing data for historical or institutional reasons even when it would be rational to do so.

Unfettered public access may be more appropriate where certain fact patterns are present. Although the following guidelines are not intended to be all-inclusive, access at no more than the marginal cost of distribution may be appropriate when

- *There is broad consensus that the resource is needed.* Unfettered access to government data is most fitting when there is little or no uncertainty that a particular resource is needed. The best evidence of consensus is a legislative or an administrative mandate specifying the need for the data among broad segments of society in support of social or economic objectives. Under these circumstances, the information has already been deemed of value for society through the legislative process.

- *The data are used for public research.* Often, government data are used in basic and targeted research both inside and outside the government to support public purposes. A government agency should obtain broad rights for public dissemination when the geographic data it acquires are likely to be useful in follow-on research and development activities.[8] When the broad public benefits of research are clear, it is appropriate for government data to be used for those purposes and, as such, to be provided at

---

[7]The term *open access* is sometimes used in this context. See definitions in Chapter 1, Section 1.4.

[8]Even in these cases, however, the government may not wish to purchase the data outright, so that the licensor still may be able to sell products derived from the data to others in the private sector.

the marginal cost of distribution. The *quid pro quo* should be that the results of the research also should be part of the public domain.

- *The data cannot be obtained by other means.* In rare cases, market failures produce a shortage of geographic data. Examples include, but are not limited to, (1) inability to obtain capital for exceptionally large or risky ventures, (2) low legal or technical barriers to copying, and (3) inability to persuade large numbers of dispersed actors to share information and resources.[9] Thus, for example, it may be appropriate for a government agency to acquire broad rights when the risks of developing geographic data are large and the government must guarantee substantial spending in order to induce investment. We stress that such market failures are uncommon and should not be presumed without clear and convincing evidence of the broader public, not private, interest. The cost-benefit calculus in these circumstances will be difficult, and is complicated by the need to assess how benefits will be shared. Government should be loathe to fund investments that result in private monopolies or oligopolies.[10]

- *The data are required as "infrastructure" upon which other datasets or data products rely.* Some geographic data supply the "infrastructure" needed to allow the integration of data among federal, state, and local agencies or to spawn new commercial products. In such cases, it may be appropriate for the government to acquire broad redistribution rights. A previous National Research Council report has suggested, for example, that environmental data should follow a "tree" model in which a government-funded

---

[9]Currently, there are indeed low technical barriers to copying geographic data, and the geographic data community has yet to develop the infrastructure and incentives that would allow efficient and effective sharing and exchange of geographic data among large numbers of dispersed actors (see Chapter 9, Section 9.3, for potential approaches for addressing these issues). However, it is difficult to argue that these low technical barriers and the less-than-optimal exchange infrastructure are so severe as to be causing shortages of geographic data in the United States.

[10]*Oligopoly* is a form of imperfect competition in which there are relatively few firms, each of which must take into account the reactions of its rivals to its own behavior (adapted from W.W. Norton and Co., 2003, *Glossary*, available at <http://www.wwnorton.com/college/econ/stiglitz/gloss.htm>).

trunk supports lush commercial branches.[11] Comparative studies show that the U.S. government's open information policies have contributed significantly to the U.S. information industry.[12] Federal government policy should continue to prevent agencies from claiming any proprietary interest in data and should continue to provide unrestricted access to public information by the commercial sector and others.

The foregoing considerations are subject to three caveats or conditions:

1. *Public provision should take place at the lowest government level that includes all potential users.* To avoid taxing one group of citizens to benefit another, geographic data should be provided by the lowest level of government that embraces all potential users.[13] For very small groups, government may not be needed at all. Instead, agencies might encourage those outside of government to pool their resources and control free riding by entering into intragroup contracts and organizations.

2. *Government should not make technical choices in anticipation of secondary and tertiary uses.* Technical specifications for government data should be based on the needs of the procuring agency and its participating stakeholders.[14] Government does not have sufficient information to choose the best solution for complex technical problems that are outside its domain. Privately held information about potential downstream uses of government data in the marketplace is best elicited and supplied by the private commercial sector.

3. *Government should not try to anticipate consumer preferences.* Government should procure data that meet its existing needs and those of its stakeholders as defined by its mandates and missions. It is difficult and often counterproductive to anticipate the number

[11]NRC, 2001, *Resolving Conflicts Arising from the Privatization of Environmental Data*, Washington, D.C., National Academies Press.

[12]See Chapter 4, Section 4.3.

[13]As noted in Chapter 6, however, it must be recognized that it may be difficult to apply this principle when, as is often the case, the potential beneficiaries cannot be identified in advance.

[14]It is entirely appropriate for an agency to develop technical specifications with its collaborators, partners, and known stakeholders.

of users and all the potential uses of a proposed dataset. Historically, government has done a poor job of determining what markets need (e.g., the Landsat program).

The foregoing considerations and caveats are offered as an aid in assessing whether to acquire data so that they may be made available to the public at or below the marginal cost of distribution. No one factor is intended to be dispositive. When multiple factors collide to generate "warning flags," alternative procurement options for acquiring geographic data should be carefully evaluated before going forward. Figure 8-1 suggests a broader framework and policy considerations for conducting this evaluation.

Recommendation: When geographic data are to be used to design or administer regulatory schemes or formulate policy, affect the rights and obligations of citizens, or have likely value for the broader society as indicated by a legislative or regulatory mandate, the agency should evaluate whether the data should be acquired under terms that permit unlimited public access or whether more limited access may suffice to support the agency's mandates and missions and the agency's actions in judicial or other review.

8.3.2.4 Limited Redistribution: The Middle Ground

When should agencies accept "reasonable" limitations on their ability to use and redistribute geographic data licensed from the private sector? Compared to other procurement methods, the costs and benefits of licensing tend to be complex. The importance of particular terms usually depends on context. Thus, there is no "golden rule" for determining which license restrictions are appropriate. That said, agencies usually need to weigh such terms as price, dissemination restrictions, available uplift rights,[15] and liability.

Recommendation: Agencies should agree to license restrictions only when doing so is consistent with their mandates, missions, and the user groups they serve.

---

[15] *Uplift rights* in a license allow future purchases by specified parties under specified terms and conditions without the need to negotiate a new license.

In some cases, potential user groups are sufficiently small for direct consultation. For example, university libraries routinely decide whether to accept particular licensing restrictions by talking to the faculty and students who are likely to use the resource. Some professors may not care if a particular license lets them reprint satellite images from an electronic journal. For others, the data may be useless without these rights. Ideally, library staff do not make such judgments; instead, they consult the affected parties directly.

Determining "reasonable" restrictions becomes harder as the number of potential users grows or is not known in advance. Researchers frequently use government data and databases to advance science. Many commercial firms use government data as a source of raw material for creating value-added industries. Citizens, educators, and scholars may derive substantial educational benefits or use government's data to check on potential abuses by government agents. These numerous other beneficiaries of government datasets will not be directly represented in licensing negotiations. Therefore, when an agency's mandates and missions specify consideration of such uses, an agency will need to consider the needs of such constituencies before licensing data subject to reuse restrictions. The needs of parties external to an agency's mandates and missions may be accommodated through government's broad information policies as specified by its laws.

As the number of potential users grows, agencies must increasingly rely on sampling to discover whether proposed restrictions are acceptable. Here, the challenge is to create an open and transparent approach that accurately conveys the user community's wishes. Many uses and users of government information are unexpected and are not likely to be identified in advance. Such needs are unlikely to be accommodated by government agents that acquire licensed data for a specific government purpose. Nonetheless, if a government agent can identify at least some government users and survey their needs, negotiations can become more informed.

Large and diverse user groups add further complications. This is because data with multiple uses typically have sharply different value for different users. Absent effective price discrimination, imposition of licensing restrictions by private firms normally will exclude at least some customers from the market. Furthermore, the number of remaining customers may not be enough to sustain the activity. Government provision may be appropriate in these circumstances.

Finally, the appropriateness of restrictions may be influenced by the nature of the content disseminated. If government users only require visualizations of datasets, rather than access to the underlying data,[16] some form of licensing restriction on the use of the underlying data may be appropriate. Conversely, license restrictions may be inappropriate when the underlying data are widely distributed and used among the government users. In this case, the data are likely to constitute part of the "public record" on which government decisions are based. If the data constitute a public record, the public must have access. Licenses that provide open access to the public, yet restrict the public from using these data in further products or ban further dissemination, may be appropriate under some circumstances.

Recommendation: Agencies that acquire data for redistribution should take affirmative steps to learn the needs and preferences of groups that are the intended beneficiaries of the data as defined by the mandates and missions of the agency. Agencies should avoid making technical choices in anticipation of secondary and tertiary uses or consumer preferences.

8.4 LICENSING STRATEGIES

When an agency decides to license data from the private sector, creative license design can make the difference between success and failure. This section provides a menu of licensing models for agencies to consider when licensing privately owned data and discusses strategies for government licensing of data to other groups.

8.4.1 Designing Licenses in the Public Interest

Having decided what mission goals a license needs to accomplish, agencies must select the right tool for the job. Agencies often face a tradeoff between cost and restrictions.

8.4.1.1 Price

If agencies can accomplish their mission at lower cost with greater benefits by resorting to, say, outright acquisition through an acquisition-

[16]See Chapter 3, Section 3.8.

for-hire arrangement or through in-house collection, they will do so.[17] Private data producers know this and realize that they must set license prices accordingly—even if it means, in the short run, licensing data at a loss. For this reason, an agency's ability to acquire data outright provides an effective "cap" on licensed data prices.

The price of licensed data typically is entangled with restrictions on redistribution, uses, or other license terms. In the interests of simplicity, this subsection focuses on pricing strategies that are more or less uncoupled from those issues.

- *Bulk Purchases.* Data providers are almost always willing to reduce prices to customers who promise to purchase large volumes of data. For example, the National Geospatial-intelligence Agency's (NGA's) recent Clearview Agreement cut the federal government's per-unit satellite image costs by roughly 75 percent.[18] The desirability of such arrangements depends on the agency's needs.[19]

- *High and Predictable Needs.* If the agency knows that it will have to make bulk purchases in any case, per-unit discounts are normally desirable and available. However, the duration and size of these bulk purchases requires careful judgment. On the one hand, agencies usually can obtain deep discounts by agreeing to large, multiyear contracts. On the other, such contracts can (1) reduce competition between vendors over the time frame of a particular contract and (2) lock in high prices in a falling market.[20]

---

[17]In written testimony to the committee (p. 2), the U.S. Census Bureau stated "The Census Bureau was unable to reach agreement with the mapping companies on these issues for an acceptable cost; acceptable generally means at a price lower than if the Census Bureau undertook the work itself."

[18]Gene Colabatistto, Space Imaging, personal communication Dec. 12, 2003.

[19]NASA's Science Data Buy is another example of bulk purchasing, in this case benefiting the academic community. For a full discussion, see NRC, 2002, *Toward New Partnerships in Remote Sensing: Government, the Private Sector, and Earth Science Research*, Washington D.C., National Academies Press, pp. 23–27.

[20]In principle, at least, both of these could be anticipated in the initial license. The risk of a multiyear contract locking in pricing above market price is particularly high now, given advances in technology and means and mechanisms of data capture. For example, direct-to-digital airborne orthophotography capture is showing cost savings of as much as 60 percent over traditional photogrammetric techniques. To date, direct-to-digital orthophotography has not been widely

Furthermore, budget uncertainties and electoral cycles can make it difficult for agencies to commit to multiyear contracts. The Nextview contract between Digital Globe and NGA is a case in point. This multiyear contract has a spending ceiling of $500 million, but it contains no spending guarantees beyond the current budget year.

- *High and Unpredictable Needs*. Agencies can find it hard to predict future need. For them, bulk-purchase contracts trade increased risk of changing needs or dropping prices for a short-term price reduction. Such trades may be sensible when vendors are willing to offer significantly lower prices. However, there is the temptation to claim credit for a highly visible price concession while downplaying the associated risk.

- *Limited Needs.* Agencies with low individual needs have a powerful incentive to band together. Indeed, some agencies collaborate because their needs are not unique enough or critical enough to justify the cost of acting independently. Various coordination strategies exist, including uplift fees, consortia, and lead agency agreements. Agency interest in better coordination within and between different levels of government is apparent from recent federal initiatives such as Geospatial One-Stop, *The National Map*, MAF/TIGER (Master Address File/Topologically Integrated Geographic Encoding and Referencing system) modernization, and the civil agency response to the U.S. Commercial Remote Sensing Policy.

8.4.1.2 Nonmonetary Factors

Over the past year, several data vendors (e.g., Space Imaging, Digital Globe, SPOT) have renewed their emphasis on long-term, bulk-sales agreements. Such risk management is particularly important to satellite data vendors, who face high fixed costs and an uncertain market. Government agencies also are interested in minimizing the risk of changing circumstances. Agencies can use the following techniques to manage risk effectively.

---

accepted. However, the technology is improving and represents a massive potential retooling in the industry, hence the risk of multiyear contracts.

- *Noncash Assets.* Agencies may find noncash assets that they can bring to license negotiations to lower upfront costs or manage risks. For example, local governments often trade government data for access to private data. Provided that such data trades do not interfere with government's dissemination mission, the practice may be commendable.

- *Pooled Resources.* Agencies and private companies often pool their resources to fund projects that neither could accomplish separately. Joint development projects and CRADAs are usually acceptable to both parties when the private partner agrees to dedicate any discoveries to the public domain. However, private partners may demand restrictive licensing terms for their discoveries or products. In this case, it may be hard to determine whether the benefits of research outweigh the "deadweight loss" associated with these terms. The negotiation may become even more difficult if the private partner agrees to share royalties with the agency. Under these circumstances, the agency may become more interested in maximizing revenues than distributing the resource at a price that encourages widespread dissemination. A preferable alternative may be for agencies to decline offers to share royalties in return for more liberal use or dissemination rights.

- *Data Certification.* Some vendors view government certification as a potential selling point.[21] In principle, agencies that invest the time and effort to provide such certification could demand a fee or discount in exchange. However, there is a sharp distinction between mandatory and voluntary government certification. Given the sophistication of most consumers, a mandatory requirement that private companies use only data certified by government (e.g., as a requirement for federal funding) could be perceived as an unreasonable infringement on consumer sovereignty. Voluntary certification would be fundamentally different because it offers consumers additional information (i.e., data screening by a trusted agency) without limiting their ability to choose uncertified products. However, certification by government that the data are accurate, complete, or fit for a particular purpose may raise liability concerns for government.

---

[21]Chapter 4, Section 4.4.

- *Network Externalities.* Information markets are said to exhibit "network externalities" in which otherwise identical goods become more valuable if they share a common standard. In the geospatial community, commercial products gain value by being compatible with government maps and digital geographic data. In recent years, local governments and commercial mapmakers and data providers have donated data to federal agencies (e.g., the U.S. Census Bureau and FEMA) to keep their products compatible. In these circumstances, the government obtained valuable data at a low or zero cost.

8.4.2 Restrictions on Dissemination

Vendors may offer federal agencies deep discounts compared to other consumers. Often, such "price discrimination" is acceptable and may even promote public policy. However, price discrimination can rarely be implemented without restricting the agency's ability to disseminate the data to secondary and tertiary users. Whether such restrictions are appropriate depends on the facts of each case. In some cases, price discrimination can be accomplished without any restrictions. In other cases, stringent restrictions may be required.

8.4.2.1 Making Licensing Choices

When agencies contemplate acquiring data with limited redissemination rights, they should ask their agency users and other likely government users which restrictions will eliminate their ability to use the data for intended purposes, which restrictions they can live with, and to what degree they are willing to make price tradeoffs. When users are relatively homogeneous, bright-line restrictions[22] offer the greatest potential for predictable outcomes. Examples include fixed update and embargo periods, blanket permission to publish isolated images or other data in academic journals, and numerical restrictions on resolution. These examples are all cases in which the benefits of licensing may exceed the costs.

---

[22]A *bright-line definition* is one that is easy and unambiguous to apply. Measurable quantities—7.2 meters, 1,000 Angstroms—are an example. In addition, the term usually connotes a willingness to accept some wrong results in the interests of having a clear, enforceable, easy-to-administer standard.

Bright-line restrictions become progressively less useful as the group of users that is covered by a license becomes larger and more diverse. Controversy often centers on users' rights to create and distribute derivative products. Most commercial satellite companies have adopted a bright-line solution that allows licensees to create and distribute any derivative product that cannot be inverted to recover the original data. Although this approach may not be practical with all other forms of geographic data and the derivative products created from them, there are instances in which this approach might be used productively with other forms of geographic imagery, such as with photogrammetric imagery.

As in any negotiation, vendors typically ask for more restrictions than their bottom line requires. For this reason, agencies should be willing to engage in negotiation. Agencies should be particularly careful not to accept restrictions unless they serve a clear and articulated business function of the vendor. For example, "best-efforts" clauses should be approached with caution.[23] On the one hand, vendors probably assign little value to such assurances; on the other, agencies are already reluctant to exercise ambiguous contract rights. Best-efforts clauses will only exacerbate this problem.

8.4.2.2 Purchasing Adequate Uplift Rights

Uplift provisions in a license allow future acquisitions by specified parties under specified terms and conditions without the need to negotiate a new license. Often, the additional parties are other government agencies. However, the government also might consider acquiring such rights in order to permit public access to the data. Agencies seldom if ever negotiate uplift rights for individual members of the public. Yet, agencies sometimes perform geographic information system (GIS) services for individual consumers on an on-demand basis. Uplift rights could be a cost-effective way to serve the relatively small number of citizens that make such requests. In many cases, agencies may be able to estimate the amount of data the public will request even though they do not know which individuals will make requests. In this case, it might be useful to prepay for the right to distribute limited quantities of data, unless the number of expected requesters or other factors weigh in favor of purchasing the data.

---

[23] *Best-efforts clauses* say that the licensee will use its best effort to exercise its contract rights in ways that preserve the licensor's ability to earn revenues from additional licenses or sales.

Redistribution rights also can be made contingent on future events. Disaster assistance is one example, and a clear list of permitted uses is important under these circumstances.[24] In principle, the ex ante price of disaster assistance data should be smaller than the ex post price, when agencies usually are prepared to pay a premium for timely information. In practice, the effect is usually small; most data vendors respond to civic demands and are supportive in emergencies.

These considerations must be weighed against the added cost of obtaining such rights in the first place. Like any other option, uplift rights add to the cost of the agency's original contract. In general, uplift rights probably do not make sense for small data procurements that are unlikely to be requested a second time.

Nevertheless, uplift rights may be underused. Negotiators who obtain uplift rights must pay higher prices and incur additional negotiating expense. When the rights are exercised, however, benefits may flow to other agencies. Well-designed uplift rights remedy this problem by providing discounts or rebates to the original agency if and when the rights are executed.

8.4.2.3 Are Restrictions Necessary?

Restrictions on reuse and redistribution are not necessary in some markets. Such markets are usually "thin" in the sense that customers have little chance of organizing an aftermarket in the vendor's data. This is common with aerial photography, where customers traditionally have received unlimited use and redistribution rights through professional data acquisition services contracts.[25] In theory, such arrangements should act as a brake on the vendor's ability to resell the data a second time. In practice, the effect is negligible. The October 1998 version of Canada's Radarsat form contract[26] provided an example of this phenomenon. It authorized licensees to create and redistribute any value-added product, including those that could be inverted to recover the original dataset. Because users had to invest substantial resources to create derivative

---

[24]See, for example, USGS Policy 01-NMD001 (April 2001): "Whenever possible, agreements should also allow the unrestricted use of such data for disaster response, research, or educational purposes."

[25]Paper maps are another example of a product that is seldom, if ever, licensed, even though they may be copyrighted. In this instance, copyright and high copying costs provide an effective business model.

[26]See Chapter 4, Section 4.3.

products in the first place, the model still provided important barriers against free riding. In effect, the licensor made a practical judgment that it could tolerate whatever "leakage" occurred. The deeper message is that contract drafters should pay more attention to what customers *will* do than to what they *might* do. The fact that data vendors seldom enforce existing contract terms[27] suggests that the earlier Radarsat-type provisions may be feasible in some other contexts.[28]

The need for use and redistribution restrictions is also reduced by large data acquisitions. In particular, such agreements (1) reduce exposure to leakage by reducing the size of any remaining market, (2) limit the risk that a vendor will lose its investment, and (3) increase the probability of leakage no matter what contract is agreed. Similarly, time-sensitive geographic data may not need use and dissemination restrictions beyond the initial embargo period, since only the up-to-date data have commercial value. Under these circumstances, vendors can afford to grant generous use and redistribution rights.

In the networked economy, companies have developed a variety of business models that facilitate widespread access to government and commercial geographic data. Many of these "New Economy" business models do not depend on royalties, minimize use and redistribution restrictions, and were built partially from, or enhance access to, government geographic data while generating revenue for the private sector:

- *Advertising.* Vendors may offer "free" services in order to support advertising or sale of related products. For example, MapQuest's Web site generates custom driving directions in order to attract viewers to Yahoo's advertising.

- *Bundling with Protected Content.* Vendors may bundle "free" data with proprietary software. For example, Caliper includes data with its proprietary GIS software.

- *One-Stop Shopping and Indexes.* Vendors may attract others to their Web site by offering indexes and links to public and private datasets. Royalties also may be collected if and when the links

[27]Many of the large data vendors who appeared before the committee remarked that they had never tried to enforce license terms or other contract rights against violators (Chapter 4, Section 4.4.1).

[28]This also suggests that accepting more restrictive license terms is unlikely to result in much of a price reduction.

generate sales (e.g., Environmental Systems Research Institute, Inc.'s Geography Network).

- *Delivery and Convenience.* Vendors may repackage public data into more convenient and more powerful formats. During the early 1990s, Warren Glimpse built a successful business transferring publicly available TIGER/Line files to CD-ROM. The TerraServer online library of satellite imagery provides a more recent example.

8.4.2.4 Government Accountability

Some agencies need broader public dissemination rights than others in responding to their mandates and missions. However, all agencies must ensure government accountability. This basic requirement sets fundamental limits on agencies' ability to accept restrictions on many types of geographic data. Fortunately, agency data acquisition and dissemination policies often are more concerned with accomplishing their core mandates and missions (including government accountability) than with economic efficiency.[29] Fortunately, there are sometimes ways to work within such constraints. One common method is to allow users to examine, process, download, and/or print views without providing access to the underlying data. The following are examples of facilities in which users have little or no need to redisseminate underlying data:

- *Libraries.* There are 1,297 libraries throughout the United States that provide no-fee access to government information in a partnership with the federal government. Many of these federal depository libraries (FDLs) provide a natural home for access to geographic information—often local, state, and federal information. During the paper map era, FDL program members ensured that government documents and data were available to members of the public. This capability was strengthened during the 1990s when many FDLs invested in onsite GIS workstations. Similarly, some state legislatures set up kiosks to provide public access to redistricting data.

- *Intranet Distribution.* University libraries may negotiate contracts that allow them to share licensed data throughout their host institutions or within multiuniversity consortia. Members of the

---

[29]Examples include redistricting data and environmental impact studies.

university community often have access to the digital information resources day or night without leaving their offices or dorm rooms and even while traveling. Typically, licensing terms limit access to these electronic resources when outside the walls of the physical library to those who are formally affiliated with the academic institution or are consortium members.

- *Access to Derived Map Views.* Some databases make selected data available for viewing and, in some cases, queries. Although they do not satisfy all possible needs, such databases may provide enough functionality, scope, and extent to meet basic demand while honoring the restrictions built into many license agreements. For example, the current implementation of *The National Map* lets anyone with Web access view detailed public geographic information about several cooperating local communities (e.g., Mecklenburg and Denver). Although online users can view and print combinations of detailed local information, they cannot download the underlying local government databases. MapQuest's Web site is an example from the private sector in which views of the results of route-planning processing are made available but the underlying geographic data remain inaccessible.

Though often acceptable, these restrictions place severe limits on users' ability to access and use data. For this reason, they normally should be thought of as setting a minimum standard of access that any government contract should meet or exceed,[30] and they will not suffice in all instances. This is particularly true when the agency's analysis of underlying data is at issue.

Although specific views may meet some public accountability needs, there are many cases in which agencies need to disclose the complete underlying data used to make government decisions. For example, views of data may be biased by those who compiled them or created the view generation software. The result may be computerized gerrymandering. Although the public does not have a right to proprietary data under FOIA, courts may not uphold agency decisions if underlying datasets are not subject to detailed public scrutiny.

[30]For example, Ernest Baldwin, U.S. Government Printing Office, testified to the committee that "at a minimum, users should be able to access and download the geographic data for reuse, at no charge, in a federal depository library."

8.4.3 The Agency as Licensor

Government mandates often may require or be best supported by free distribution of acquired data. At the federal level, licensing of government data to others typically occurs through specific legislative exceptions (e.g., CRADAs,[31] Landsat Commercialization Act), and when there is a need to "pass through" commercial license provisions that apply to proprietary data that the government possesses (e.g., NGA's Clearview Agreement and data from the USGS's SPOT data archives in Sioux Falls). Otherwise, federal agencies are severely constrained in how they may limit access to data in which they possess full rights. The same also applies to many—though not all—state and local government agencies. These restrictions usually limit the ability of federal agencies and many local agencies to recover fees above the marginal cost of distribution.

The next subsection begins by discussing the revenue generation and pricing goals that an agency choosing to distribute data under license to outside users might pursue. The second subsection addresses how agencies can use licenses to achieve various noneconomic goals.

8.4.3.1 Revenue Generation and Pricing Goals

During the 1990s, many jurisdictions experimented with cost-recovery fees for geographic data. Ten years later, many of these entities have concluded that fee programs[32]

- cannot recover a worthwhile fraction of government data budgets,
- seldom cover operating expenses, and
- act as a drag on private-sector investments that otherwise would add to the tax base and grow the economy.

---

[31]CRADAs involving digital geographic data may or may not impose restrictions on the dissemination and use of the data produced or made more readily accessible through the CRADA. See the CRADA involving the National Oceanic and Atmospheric Administration's hydrographic charts and the USGS-Microsoft TerraServer CRADA described in Chapter 4, Section 4.2.2.

[32]See Chapter 4 (Section 4.3).

Recent blue-ribbon reports[33] and statements before this committee[34] suggest an emerging belief that cost-recovery experiments with public geographic data in the United States have failed to generate significant revenues while meeting the functions of government. This result may have been predictable because the demand for geographic data is particularly price sensitive. Many local government agency experiments in cost recovery sought prices beyond what most of the potential market was willing to pay, and so, buyers chose to do without the data or found substitute data. Local governments often found it cumbersome or politically impractical to embrace the price discrimination and other market segmentation strategies that their commercial competitors adopted. In addition, there is a limited market for local geographic data.

Nevertheless, state and local government agencies may choose to charge a higher fee,[35] or local law may require them to set prices according to a particular standard (e.g., cost recovery). When local government jurisdictions *choose* to distribute geographic data through sale or license, agencies could limit the impact of such a decision by adopting the following strategies:

- *Price Discrimination.* Price discrimination mitigates the economic inefficiencies associated with user fees.[36] Agencies should resist calls to make data available to all clients at a single price.

- *Constrained Optimization.* There are distinctions between (1) setting prices at higher than the marginal cost of distribution with or without regard to the revenue stream actually generated, (2) attempting to cover ongoing costs, and (3) attempting to become a "profit center." If the goal is to finance ongoing operations while still providing affordable public access, agencies should charge the lowest price consistent with covering their variable costs. Agencies should resist the urge to set high prices if that results in the exclusion of a significant number of users.

---

[33]See Chapter 4, footnote 95.

[34]See, for example, testimony of Bob Amos, William Burgess, Randy Johnson, and Peter Weiss.

[35]See Open Data Consortium [ODC], 2003, *10 Ways to Support Your GIS Without Selling Data*. Available at <http://www.opendataconsortium.org> (hereinafter ODC, 2003).

[36]See Chapter 6.

- *Minimally Restrictive Contracts.* Previous sections have discussed a broad menu of licensing terms that agencies can demand from vendors. Agencies choosing to distribute data under license should offer similar terms. The October 1998 version of Canada's Radarsat license provides an example of minimally restrictive government data licensing. This license relaxes constraints on inverting data back to their original form. The license design takes into account what users *will* do rather than what they *might* do.

8.4.3.2 Nonfinancial Reasons for Licensing by Government

Even when cost recovery is not a goal, agencies may sometimes use licenses to pursue other, nonfinancial policy goals. Such goals include attribution, negating implied endorsements, and risk management. Like any other license terms, these provisions should impose minimal restrictions on licensees' ability to use and redisseminate data.

- *Attribution.* Twenty years ago, it was easy for an author to cite each and every source consulted. Today, data products frequently extract, combine, and modify millions of data points from dozens of sources. In this new context, it may not be reasonable—or even feasible—to require users to provide an individualized attribution "tag" for each piece of data. Unless technological advances provide a solution, a simple statement that "This product contains data originally gathered and compiled by Agency XX" may have to suffice.

- *Negating Implied Endorsements.* Geographic data are most valuable when they can be combined and repackaged to create new products. Agencies have a legitimate interest in reminding the public that these products are not endorsed or certified by the government. However, it is economically counterproductive for agencies to accomplish this goal by banning the extraction and modification of data altogether. A simple, prominent disclaimer is usually enough to negate any implication of endorsement or warranty.

- *Managing Risk.* Indemnity and liability disclaimers are often reasonable and should be encouraged. Agencies are understandably reluctant to distribute data if doing so exposes them to liability. This is particularly true when data are distributed at or near marginal cost. Disclaimers may legitimately extend to tertiary users, particularly if made as an explicit condition of the licensing arrangement with the secondary user. Well-designed disclaimers have little or no impact on consumers' ability to extract, use, or manipulate data.

8.5 ACCOMMODATING A "CULTURE OF LICENSING"

Most data vendors' terms and prices are negotiable, particularly for large transactions. Agencies with narrow, clearly defined projects rarely need unfettered rights in geographic data and may decide that it makes both short- and long-term sense to demand fewer rights in exchange for lower prices. Agencies also may be able to offer in-kind payments to vendors to lower dollar costs still further. These include, but are not limited to, such resources as publicity, access to agency expertise, and data verification.

The committee heard many examples of agencies that manage to negotiate favorable and often innovative contracts. That said, some agencies seem to believe that they cannot negotiate from a position of strength or else find negotiations burdensome. As a result, some agencies indicated that they accepted vendors' opening offers at face value with little or no negotiation. Not coincidentally, these same agencies tended to have the most disappointing licensing experiences.

Finally, some agencies report that "uncertainties" in past licenses have deterred them from worthwhile projects. Yet they did not contact the vendor, much less demand that the uncertainties be resolved in their favor. The culture of licensing does not end once a contract is signed. Asserting contractual rights is a necessary part of living with licenses.

Conclusion: Given the expansion of licensing of geographic data in the marketplace, agencies cannot help becoming more sophisticated consumers when licensing is the only or best-value option in acquiring geographic data. Qualifications-based selection procurement accompanied by subsequent cost negotiations and, when appropriate, traditional competitive bidding practices can help agencies obtain the best possible terms.

Recommendation: Agencies should dedicate resources to training and knowledge-sharing among agencies in order to extract maximum public benefit from licensing. The Federal Geographic Data Committee's working group and subcommittee structure provides a

convenient venue through which agencies can report and learn from their experiences.

VIGNETTE H. THE SPATIAL SEMANTIC WEB DREAM

"With the growth of the World Wide Web has come the insight that currently available methods for finding and using information on the Web are often insufficient. ...Today's retrieval methods typically are limited to keyword searches or matches of substrings, offering no support for any deeper structures that might lie hidden in the data or that people typically use to reason; therefore, users often may miss critical information when searching the Web. ...The advent of the Semantic Web promises better retrieval methods by incorporating the data's semantics and exploiting the semantics during the search process."[37]

Dianne Hamilton has created a land-ownership-parcel dataset and would like to make it known and available to the world. As she completes a minimal set of online metadata questions, she is asked to supply the "type of dataset." She selects "parcel" from a pull-down menu. The system responds by asking her to select from several definitions of parcel or to construct her own definition. She clicks the supplied definition of "land ownership parcel" (as opposed to "package parcel" or "land use parcel") and the remainder of the metadata entry process becomes simpler because the top items in the pull-down menus for subsequent entries are options that most closely comport with land ownership ontologies.[38]

Ms. Hamilton's choices in her remaining entries automatically tie her data to one or several ontologies, thereby enabling semantic search engines to find her dataset using criteria based on her supplied deeper understanding of the content. Although Ms. Hamilton, a novice dataset creator and publisher, will not know and cannot supply the technical description and details of models used to construct her data product, the process she has followed nevertheless ensures that such details can be incorporated within and later automatically extracted from her published dataset.

For the Spatial Semantic Web to reach its full potential, automated searches must be able to reach and explore actual geographic datasets as well as their metadata. Without full access to the dataset, data semantics cannot be used to find and assess the suitability of Ms. Hamilton's data for an explicit need. Additionally, searches that rely on data similarity assessments require access to the data rather than just metadata.

---

[37]M. J., Egenhofer, 2002, Toward the Semantic Geospatial Web, ACM GIS 2002 (Nov. 8–9), McLean, VA, Association for Computer Machinery.

[38]An *ontology* is an explicit formal specification of how to represent objects, concepts, and other entities that are assumed to exist in some area of interest, and the relationships among them.

In the past, Ms. Hamilton would never have made her dataset fully available on the Web because it was so easy for others to simply take it. However, because she can now readily incorporate standard license language in the metadata and identifiers in the geographic data, she has developed confidence in her ability to track and substantially reduce such free riding. By taking this legal and technological approach, Ms. Hamilton helps the Spatial Semantic Web reach its greatest functionality and thereby enhances discovery and usage of her offerings.

In the end, the Spatial Semantic Web dream comes down to this: Can a combined technological and licensing infrastructure be developed that supports easy and efficient online entry of licensing and technical information?

9

New Institutions

9.1 INTRODUCTION

This chapter presents interventions, strategies, and models for new institutions that could make licensing a more powerful and attractive tool for government agencies, commercial firms, and other affected stakeholders.[1] The institutions we propose follow naturally from the problems, experiences, and analyses presented in earlier chapters. The vignettes between these earlier chapters provide glimpses of how future operational environments might better serve the needs of the broadest range of stakeholders in geographic data and services.

In what follows, the committee does not try to present detailed blueprints for any specific program. Instead, we describe a range of generic options. For now, community debate should focus on which options deserve to be pursued. Once these decisions have been made, detailed design will be needed to make the new institution maximally useful and to ensure that it balances the interests of all parties affected by licensing of geographic data and services to and from government.

The chapter is in three sections. The first focuses on the need for standard-form licensing agreements and new institutions for coordinating government acquisitions. The second explores the related concepts of a national commons and marketplace for geographic information. The third

[1]Addressing the committee's sixth task, and also aspects of its fourth and fifth tasks.

section discusses how the commons and marketplace might evolve to benefit all stakeholders.

In contrast with the preceding chapter, the implementation of most of the ideas in this chapter will take time. Building new mechanisms and institutions to make licensing more productive will require sustained initiatives by federal, state, and local agencies, and, in many cases, the private sector.[2]

9.2 STRUCTURAL INTERVENTIONS

9.2.1 Standard Licenses and Form Agreements

Geographic data contracts come in a diverse range of styles and levels of complexity. Contracts for large transactions often are negotiated "from scratch." Contracts for small transactions often use idiosyncratic "form contracts" that differ from vendor to vendor. Greater standardization could lead to reduced uncertainty in procurement, lower negotiation costs, and probably increased numbers of licenses. At a minimum, it may be feasible to standardize straightforward provisions covering liability, indemnity, attribution, jurisdiction, and choice of law.

Some standardization will emerge naturally as parties gain experience in contracting.[3] Agencies, trade associations, and public interest groups can accelerate this process by creating recommended contracts and compiling online or printed form books. Some steps already have been taken in this direction[4] and further progress is likely.

Standard language and (eventually) standard form licenses are key to many of the recommendations contained in this report. Relative to the amount of time and effort that industry expends each year in negotiating

---

[2]Some actions, including the development of model licensing agreements, could evolve more rapidly.

[3]This is not quite as easy as it sounds. Some vendors keep contracts secret in hopes of gaining a competitive advantage. Nonetheless, disclosure normally should be a matter of enlightened self-interest, and the potential benefits to the industry far outweigh competitive advantage in most cases.

[4]For example, the U.S. Geological Survey (USGS) has recently circulated a proposed Model Contract for "Purchase of Satellite Data." Public Technologies, Inc., also promotes standard contracts through its "best practices" program for local governments, and the Open Data Consortium has released a model data distribution policy (see <http://www.opendataconsortium.org>).

and interpreting contracts, the required investment is likely to be modest. Immediate and long-term benefits include

- *Reduced Negotiation Costs.* One witness reported being able to jumpstart stalled negotiations by adopting old licensing language.[5] Widespread dissemination of standard language and licenses would similarly reduce the time and expense needed to reach agreement.

- *Reduced Uncertainty.* Contracts are often ambiguous. If the drafters' intent is not obvious, often, as a last resort, judges will interpret the contract and supply meaning. Form language accelerates this process because it facilitates the development of commonly accepted interpretations. It also lets courts revisit, and resolve, points of ambiguity. In addition, standard contracts provide familiarity and certainty. Corporate counsel are more apt to approve license terms that have withstood the test of time and litigation.

- *Improved Market Efficiency.* Reduced negotiation costs and well-defined terms make markets more efficient. Form contracts advance these goals.

- *Increased Automation.* Standardized contracts lower the cost and complexity of computerization. Good form contracts are an important precondition for advanced data brokerages, business-to-business systems, and online markets.

Best-business-practice contracts eventually will find their way into the government geographic data market. Agencies can accelerate the process by encouraging organizations involved in geographic data transactions to develop, recommend, and publicize high-quality clauses and form agreements.

Recommendation: Agencies, trade associations, and public interest groups should exercise leadership in promoting standard clauses and form licenses throughout the geographic data community.

[5]Testimony of Chris Friel, GIS Solutions Inc.

9.2.2 Coordinating Government Acquisitions

Licensing can facilitate coordination of geographic data procurement by government agencies in two ways. First, agencies can agree to coordinate their acquisitions under multiagency, or umbrella, licenses. Second, government can create institutions that achieve coordination "automatically." Automation and market signals hold great promise for improving large-scale coordination within and among federal, state, and local government levels.[6]

9.2.2.1 Multiagency Licenses

Individual agencies have a strong incentive to participate in licensing arrangements that cut across traditional agency functions. In most cases, the rewards are bulk discounts and shared transaction costs. Nonetheless, the practicality of this model depends on the facts of each case. For example, the benefits of collaboration, although strong, are sometimes dismissed on grounds that an agency's needs are "unique." Coordination also may be too costly for small transactions or may become increasingly difficult as the number of agencies grows. In general, agencies face four options when considering procurement under license:

1. *Individual Procurement.* This option is sometimes rational, particularly for small, one-off acquisitions when individual agencies find it prohibitively expensive to (a) anticipate all future uses throughout government, (b) identify each user, (c) determine the needs and preferences of these users, and (d) negotiate and administer a governmentwide contract. Although agencies have a built-in incentive to discount or ignore benefits that accrue to other agencies or parties, they should work to overcome this bias where appropriate to support broader government goals.

2. *Click-wrap or Shrink-wrap Licenses.* Mass-market, low-cost data products are the strongest candidates for uncoordinated procurement. Such products typically are bundled with shrink-wrap or click-wrap licenses that limit the customer's right to use and redistribute data.[7] Although agencies could theoretically band

[6]Multiagency and automated procurement by purchase, as opposed to licensing, may result in some or many of the same benefits.

[7]The enforceability of these clauses is unclear (see Chapter 5, Section 5.3.1).

together to obtain better terms, the required transaction costs likely would exceed any benefits,[8] particularly when an agency's mission limits data usage to internal use.

3. *Uplift Licenses.* Transaction costs can be high when individual negotiation is required. However, the incremental expense of negotiating uplift rights for the potential benefit of other agencies on the same or similar terms is usually low. Government has made progress in using uplift rights to achieve coordination across agencies.[9]

4. *Coordinated Acquisitions.* Consortia of interested government entities—sometimes known as cooperative funding partnerships—may be organized to bargain with vendors.[10] Because consortia are voluntary, members can walk away at any time, making most consortia highly responsive to member needs. Assuming that the needs of members are not too diverse, consortia often can be unified sufficiently to be effective negotiators.

Most federal agencies take an approach in which a single "lead agency" represents multiple users across multiple agencies. This hierarchical strategy has benefits and costs compared to consortia. On the benefits side, vesting discretion in a lead agency minimizes the need for ongoing interagency meetings and interactions. Furthermore, lead agencies typically possess above-average technology and licensing expertise and will acquire additional expertise by negotiating on behalf of others.[11] On

---

[8]Transaction costs include, but are not limited to, time spent on interagency meetings, negotiations with vendors, review by government lawyers, and the logistics of copying and distributing data to any employee who requests it.

[9]See, for example, USGS Policy 01-NMD001 (April 2001) ("Procurement contracts should also contain terms to allow additional rights to be purchased"). The National Oceanic and Atmospheric Administration (NOAA) has been particularly active in organizing uplift rights around specific projects and geographic areas (e.g., the NOAA/Intermap Santa Cruz/San Mateo County digital elevation model license that provided uplift rights for the Federal Emergency Management Agency (FEMA), USGS, and a private partner). The National Geospatial-Intelligence Agency (NGA) also uses uplift rights in contracts.

[10]See Chapter 4.

[11]For example, NGA has acquired licenses for various commercial street-centerline and fire station files on behalf of the federal government (although the license does not permit use by the U.S. Census Bureau). Similarly, USGS established a program for government acquisition of Landsat and Systeme Probatoire

the cost side, lead agencies tend to be less sensitive and less knowledgeable of the explicit needs of interested users across the agencies they represent. This lack of knowledge may limit the lead agency's ability to negotiate win-win agreements with vendors. Presumably, a lead agency also may be tempted to place its own needs ahead of others. Testimony from the USGS to the committee acknowledged both the weaknesses and potential of the lead-agency approach:

> Even with the existence of a centralized procurement mechanism [for satellite images…it was still very difficult to truly represent a single unified voice on behalf of the government during negotiations with SPOT Image Corporation. If the federal agencies could have unified their unique data requirements and associated funding into a single negotiation with SPOT, the government would have been in a much better position to negotiate licensing terms.

Federal agencies are considering initiatives that would extend multiagency procurement coordination to larger scales.[12] The extent to which governmentwide consortia can be simultaneously manageable and responsive is unclear.

9.2.2.2 *Markets and Automation*

Cooperation among government agencies acquiring licensed geographic data need not involve coordinated negotiations. An alternative is to cooperate in creating institutions that reduce the costs that each agency

---

Pour l'Observation de la Terre (SPOT) data in 1986. The acquisition licenses ranged from single-agency contracts to governmentwide contracts. Although the licenses were optional, more than 30 federal agencies have acquired over $42 million worth of images through the USGS program. Paperwork was reduced to a single purchase order. In 2001, USGS acquired full ownership of all Landsat data, which can now be shared without limitation. SPOT data are still distributed under the original arrangement.

[12]The NGA's proposed National Commercial Imagery Strategy to coordinate satellite data procurement within the defense community was followed by USGS's suggestion of a parallel civilian strategy as part of its National Spatial Data Infrastructure (NSDI) Initiative. Some proponents hope to merge both initiatives within a single "National Strategy." At the same time, proponents realize that genuinely integrated, nationwide procurement for federal agencies could shortchange small civilian agencies.

incurs in acquiring data.[13] We discuss two prominent candidates: data brokerages and business-to-government networks, and suggest a potentially cost-effective interim compromise (a standard license provision search capability):

- *Data Brokerages.* Government employees cannot take advantage of previously acquired government data (including uplift rights) unless they know about them and the details of limitations on use. During the 1990s, USGS launched an online "Data Brokerage" with which federal employees could search for previously licensed data. USGS ultimately abandoned the site because it was rarely used. In addition, the cost and complexity of tracking multiple, nonstandard license terms were prohibitive.

  Nevertheless, the USGS's initiative may have been ahead of its time. The basic concept is sound and could yield substantial savings for the federal government, assuming that at least some previously licensed data are relevant to current needs and the limitations on use are not onerous to understand or meet. Data brokerages will become increasingly feasible if government and industry start to use standardized licenses.

- *Business-to-Government.* USGS's data brokerage system would have required government employees to point and click their way through on-screen licenses. In principle, computers can do this job quicker and more efficiently. Many commercial companies already use business-to-business (B2B) systems to eliminate this workload. In a typical system, employees report their needs to a computer that aggregates companywide demand, procures bids, "negotiates" contracts, and pays invoices. B2B is particularly important in manufacturing (automobiles, aerospace) and retail industries (warehouse stores) where companies need to manage just-in-time inventory. Business-to-government (B2G) data licensing systems eventually could accomplish similar coordination of standard data acquisitions from competing vendors at significant savings.

  The delivery of geographic data and services through B2G systems will be increasingly important in the long run. There are already a number of B2B and B2G Web mapping services delivering standardized location services such as geocoding, route

---

[13]Government agencies are not the only potential beneficiaries of improved institutions. Individuals, businesses, and academic organizations may also benefit.

mapping, address validation, localized weather, traffic reporting, and property risk assessment.[14] Predictably, standard interfaces and protocols, along with uniform licensing forms, are emerging to reduce the cost and complexity of syndicating these services. These types of online services hold significant potential for many public-sector organizations and smaller commercial companies that cannot afford the upfront costs of hiring staff and acquiring hardware, software, and data to deploy their own service.

- *Standard License Provision Search Capability.* A compromise between data brokerages and B2G might be to develop and support a standard license provision search capability for geographic data.[15] Each time a government agency licenses data from a vendor allowing limited use by others or acquires uplift rights from a vendor, the agency would post the specific license terms (ideally reduced to a uniform code) and related metadata on its Web site.[16] In most cases, commercial search engines (e.g., Google) would digest the posting within days, eliminating the need for a central index. Thereafter, other government users could find licensed geographic data that meet their needs by using standard text Web searches. By taking advantage of commercial search engines that search for license provisions and metadata in standard forms, agencies could build a continually updated inventory of previously licensed data at little or no cost to themselves. This could also lower the cost of acquiring additional rights or data.

    A standard license provision search capability would not support B2G-style automated procurement. However, it would be much simpler to build. Furthermore—and unlike B2G—small

---

[14]See, for example, <http://www.mapquest.com/solutions/product.adp>; <http://www.microsoft.com/mappoint/webservice/default.mspx>; <http:// www.meteorologix.com/industry/homeland.cfm>; <http://www.questerra.com/platform/index.html>; <http://www.esri.com/software/arcwebservices/about/overview>.

[15]For example, the Creative Commons project supports the automated generation of standard license provisions, the posting of licenses with the digital work product, and promotes ease of Web search in support of open access sharing and use. If desired, a similar online licensing support approach and search capability might be developed for support of more restrictive licensing terms in the use of geographic data. Creative Commons concepts are described at <http://www.creativecommons.org>.

[16]Alternatively or additionally, the vendor could post this information on its Web site.

vendors could participate immediately. Finally, the capability would provide a useful stepping-stone to B2G.

Recommendation: Agencies should continue to keep abreast of data brokerage and automated purchasing system developments that might help them coordinate data acquisitions from competing vendors.

9.3 TOWARD A NATIONAL COMMONS AND MARKETPLACE

Society benefits when its members can find and use desired existing resources and products. Facilitating the sharing of and trade in data through the development of an efficient and user-friendly system, including a well-organized commons connecting users and contributors and an efficient market connecting buyers and sellers, would be a valuable endeavor.[17] Although no such online environment currently exists for geographic data, *The National Map,* Geospatial One-Stop, and the NSDI provide first steps.

In this section we describe a vision for a National Commons and Marketplace in Geographic Information.[18] In a later section we suggest how the two might be integrated and operate seamlessly, and discuss options for who should develop and host them.

9.3.1 A National Commons in Geographic Information

Commercial, scientific, and nonprofit users rely heavily on public domain geographic information to create value-added resources. Such resources can be expanded by a National Commons in Geographic Information (hereinafter "National Commons") that aids creation of public

[17]Vendors understand the value of a national market. If transaction costs—broadly defined—could be reduced, then more data would be produced and the price to any individual user would decline. One vendor told the committee that he would cut prices by three fourths in a market that let him reach agency buyers (testimony of David DeLorme).

[18]The vignettes placed between the chapters of this report provide glimpses of future capabilities that might be enabled through a combined commons and marketplace founded on licensing. The commons and marketplace concepts are introduced principally in Vignette F, "A Mainstream Geographic Data Marketplace Dream," and Vignette G, "A Global Information Commons Dream." Here, the concepts are presented in greater detail and in a national rather than global context.

domain resources and open access content and makes them readily accessible.[19]

The overarching goal of the National Commons is to create a broad and continually growing set of freely usable (i.e., no monetary charge for use) geographic data and products at local scales similar in effect to the public domain datasets and works created by federal agencies. To succeed, the commons could provide easy, effective, and integrated mechanisms that could, for example,

- enable any geographic dataset creator to construct a license that grants permission to use his or her data,
- enable novice creators to quickly generate accurate and substantive standardized metadata for a geographic data file,
- enable data contributors to take advantage of form liability disclaimers,
- embed identifiers automatically in any commons dataset so that any future user can link back and recover the detailed metadata and license conditions for the file,
- allow for deeper search capabilities of geographic data and metadata than are currently available, and
- provide a long-term archive for commons geographic datasets.

Initial components of a National Commons could be implemented almost immediately with minimal investment (e.g., the first three bulleted items might be implemented as extensions of Geospatial One-Stop and Creative Commons efforts), while the fully envisioned system appears achievable on the basis of existing knowledge.

Not all local governments, private citizens, or private companies will want to make any or all of their geographic datasets or products available in the National Commons. Nevertheless, more people will participate once a large, user-friendly capability is available. A simple user interface (see, e.g., Box 9-1) could facilitate this process.

Today's geographic data users can assume that most U.S. federal geographic datasets are available with no intellectual property limitations attached to them;[20] but this assumption is not valid for most other digital geographic information. A National Commons in Geographic Information could allow any data creator to quickly construct a comprehensive, standard, and yet flexible license granting others permission to use the

[19]See Chapter 1, Section 1.4, for definitions of public domain, open access content, and geographic information commons.

[20]See Chapter 5, Section 5.4.

creator's work. By analogy with the Creative Commons license process, creators might be offered license options to (1) allow public domain use for any purpose, (2) require attribution, (3) allow or disallow commercial uses, (4) allow or disallow modification of the work, and (5) allow modification as long as others use the identical license with their derivative works (commonly referred to as "share-alike" or "copyleft"). The commons license would also offer standard liability disclaimers—an important feature for utilitarian works such as geographic data upon which decisions are likely to be based.

Last, the commons license model[21] gives value-adders the ability to charge for the service of transferring their work to others[22] and a variety of support services.[23] However, data contributors would receive no royalties or rents from others for use of their data.

Recommendation: The geographic data community should consider a National Commons in Geographic Information where citizens can post and acquire commons-licensed geographic data. The proposed facility would make it easier for geographic data creators (including local to federal agencies) to document, license, and deliver their datasets to a common shared pool, and also would help the broader

---

[21]The commons license model has objectives similar to those of licenses used in related open source and open access initiatives. Apache, Linux, Perl, and Sendmail are examples of widely used software developed through distributed contributors adhering to open source licenses. Examples of collaboratively produced open access information works may be found at The Directory of Open Access Journals (<http://www.doaj.org>), UNESCO Social Science Online Publications (<http://www.unesco.org/shs/shsdc/journals/shsjournals.html>), Wikipedia (<http://www.wikipedia.org>), the Open Textbook Project (<http://www.otp.inlimine.org>), and the Gutenberg Project (<http://www.gutenberg.net>). For a sampling of open source software and open access products directly germane to the geographic information community, consult <http://freeGIS.org>. The use and sharing expectations in most of these open collaborative efforts are defined by explicit licenses or published policies.

[22]For example, charges are made for downloading the open source movie clips found at <http://nothingsostrange.com/open_source>. Those who pay the fee to download are free to use, copy, and disseminate the clips as well as use them in other commercial and noncommercial derivative creations, provided that attribution is given.

[23]Redhat (<http://www.redhat.com>) is a corporation that generates income by delivering recommended open source software and providing professional services, technical support, and training in the use of such software.

community to find, acquire, and use such data. Participation would be voluntary.

Box 9-1
Conceptual Model for a National Commons in Geographic Information:
Possible Operational Characteristics[a]

1. A nonexpert user creates a geographic dataset that she or he wants preserved and accessible to the rest of the world.
2. The user accesses a Web site that automatically generates a commons license and facilitates the creation of a metadata record in response to a Web interview.[b]

(a) *Commons License Creation.* In responding to the interview, the contributor either dedicates the file to the public domain or chooses among a limited selection of "commons" license provisions to apply to the dataset. The basic goal in this instance is to allow the data file creator to notify subsequent users that they may use the file without asking for permission under wide-ranging conditions at no monetary charge. Possible limitations imposed by the data creator may include (i) requiring users to provide attribution, (ii) disallowance of modifications, (iii) disallowance of commercial use, and (iv) liability disclaimers.

(b) *Metadata Creation.* The user is asked about the details of the dataset through a series of plain English questions and limited-choice responses. The system guides the user to provide deeper meaning to the selected pull-down descriptors by asking the contributor to pick among definitions for the metadata the contributor has selected. Those definitions along with formal specifications for the potential domains of interest (i.e., ontologies) are used to predict and simplify subsequent metadata selection choices. Open-ended questions with free-form responses are minimized and metadata fields are automatically populated whenever possible. Despite being invisible to most users, the resulting metadata permit far more nuanced and accurate searches than current technology.

3. The interview responses and the accompanying geographic data file are submitted to an automated processing facility. An encrypted identifier is embedded in the file but does not interfere with it.[c] The identifier cannot be stripped from the file through standard geographic information system (GIS) operations,[d] and may be linked back to the full commons license and metadata at any time over the Web. Through freely down-

loadable client software, any user may readily determine the status of legal rights and metadata for any standard-format geographic data file they possess. The originator and the string of value-adders are readily identified from a file processed in this manner. The existence of identifier information in a file is also strong evidence that the owner has authorized its use. Would-be infringers who attempt to remove or alter metadata information cannot be certain that additional, undetected identifiers do not remain hidden in the file.

4. The system returns a copy of the "marked" geographic data file incorporating the embedded license and metadata link to the originator. The creator also can choose to have the file centrally and openly archived. Archiving ensures a backup for commons-licensed data files that would otherwise be distributed among thousands of computers, inevitably giving rise to broken links and lost data. If archived, the system may generate and make accessible several standard and interchange formats for the data file.[e] Whether the data are maintained on the open Web or in a long-term electronic archive, potential users can search for, access, and download such datasets.

5. A capability is provided for user or peer assessment of the quality and usefulness of the supplied metadata as well as the geographic data files.[f] The system also provides a means for reaching people interested in using or contributing commons-licensed geographic data.[g]

---

[a]See <http://www.spatial.maine.edu/geodatacommons> for an example of a Web mockup illustrating steps described in this box. See also Vignette G in this volume.

[b]Alternatively, the commons could offer downloadable software to accomplish these tasks.

[c]This intellectual property management system begins with an assumption of open access by all to the datasets as opposed to more traditional digital rights management architectures that begin with the assumption that only users with authorization should be granted access. The unique identifier may or may not be a "hash," that is, an identifier based on the digital file's contents.

[d]For raster files, several means for embedding such an identifier have been developed. For vector files, see W. Huber, 2002, Vector steganography: A practical introduction, *Directions Magazine* (April 18), available at <http://www.directionsmag.com/article.php?article_id=195>.

[e]For example, Citeseer (<http://citeseer.ist.psu.edu> or <http://www.citeseer.com>) finds articles on the Web and downloads them in existing formats but provides users with several standard versions of the same file.

[f]For example, see the rating system supported by slashdot.org as discussed by S. Johnson, 2001, *Emergence: The Connected Lives of Ants, Brains, Cities and Software*, New York, Scribner, pp. 152–162; and the rating system used by eBay (see <http://www.ebay.com>).

[g]For example, see the discussion by A.-L. Barabasi, 2002, *Linked: The New Science of Networks,* Cambridge, Perseus, pp. 213–216.

9.3.2 A National Marketplace in Geographic Information

The Internet has enhanced the ability of commercial businesses, government, nonprofit organizations, and individuals to find geographic data that meet their needs. Commercial vendors make their data offerings known and available through corporate Web sites, online index sites, and portals. Yet, as discussed in previous chapters, the ability to find, assess, and acquire these data could be far more efficient—to the benefit of all. A national marketplace in geographic information would provide an online environment where any seller or licensor, no matter how small, could efficiently post its geographic data offerings in a searchable form using a menu of standard license choices and metadata reporting. Would-be customers could search through the thousands of data offerings, select the product that meets their technical and license condition needs, perform efficient comparison shopping, and buy or license the desired geographic data file within minutes of finding it.

In the simplest implementation of the marketplace, customers would obtain the data directly from the vendor after "clicking through" to contact its server. In more advanced implementations, the seller or licensor might define for each dataset or group of datasets a pricing formula that varies with differing standard license or sale conditions. Advanced systems could also provide automated financial transactions and product delivery. Sellers' accounts could be automatically credited with funds from direct or downstream derivative product sales. Sellers could alter their geographic data offerings, descriptions, license conditions, and pricing formulas at any time.

Recommendation: The geographic data community should consider a national marketplace in geographic information where individuals can offer and acquire commercial geographic data. The proposed facility would make it easier for the geographic data community to offer, find, acquire, and use existing geographic data under license. Participation would be voluntary.

The marketplace vision assumes that even a small company whose core business does not involve selling geographic data can participate. Simple, user-friendly systems (see, e.g., Box 9-2) will be the key to extending the marketplace to the largest possible number of buyers and sellers. The immense variety of traded geographic data and services is dealt with in this marketplace vision by standardizing the documentation of such data and services and making that documentation searchable efficiently and effectively.

So far, we have discussed the commons and the marketplace as if they were separate institutions. In practice, the software and hardware needed to build both projects would be very similar. For this reason, it would make sense to pursue both concepts simultaneously—as a single facility or closely integrated facilities. Most consumers do not care whether they use “public domain,” “commons,” or “commercial” data, provided they are able to find information meeting their desired technical requirements, use conditions, and costs.

Box 9-2
Conceptual Model for a National Marketplace in Geographic Information: Possible Operational Characteristics[a]

1. A commercial company has created a geographic dataset that it wants to offer to potential purchasers or licensees.

2. The company accesses a Web site that performs the following functions:

(a) *License Term and Pricing Definition.* The Web site automatically offers a wide variety of standard license provisions. In response to a transcript, the data supplier mixes and matches desired standard license provisions, and is led through a process for standardized posting of its price schedule.
(b) *Metadata Creation.* Next, if standardized metadata have not already been created, the program asks the supplier to describe their data through a series of questions with limited-choice responses. The system uses pull-down descriptors that aid the user in providing detailed information.[b]

3. The transcript responses and the accompanying data file are submitted to an automated processing facility. An encrypted identifier is embedded in the file but does not interfere with it. The identifier cannot be stripped from the file through standard GIS operations, and may be linked back to the full commercial license and metadata at any time over the Web. This allows users to readily assess whether the data will meet their needs, and the vendor can affirmatively notify all potential subsequent users of the legal uses that may be made of the data file.[c]
4. The system returns a copy of the "marked" geographic data file to the originator, incorporating the embedded link to the license and metadata. The marking process increases the efficiency of automated data searches by consumers with specific needs and conditions.

5. The system allows users to post comments on the quality and usefulness of each file's content as well as on the metadata. It also provides a means for reaching people interested in finding or offering data.[d]

---

[a]See also Vignette F in this volume. For a summary of alternative or additional digital rights management architectures that might be explored for geographic data, see R. Iannella, 2001, Digital rights management (DRM) architectures, available at: <http://www.dlib.org/dlib/june01/iannella/ 06iannella.html>. Such architectures typically begin with the assumption that only authorized users will be allowed access to the data.

[b]See step 2(b) of Box 9-1 for a parallel but more detailed discussion.

[c]See step 3 of Box 9-1 for a parallel but more detailed discussion.

[d]See step 5 of Box 9-1 for a parallel but more detailed discussion.

9.4 POLICY CHOICES

A properly designed and integrated online National Commons and Marketplace in Geographic Information (hereinafter, the National Commons and Marketplace) could make agency licensing more efficient, reduce wasteful duplication between agencies, accelerate the availability of local datasets in the public domain and commons, improve archiving of geographic data, increase the range of geographic data products available to consumers, and foster competition among private vendors. However, these outcomes are far from inevitable. Absent strong agency leadership, the institutions that actually emerge may offer fewer benefits.

A National Commons and Marketplace might be operated by government, the private sector, or through a division of responsibilities between them. Privately owned institutions pose significant antitrust concerns.

9.4.1 Government as Operator

The National Commons and Marketplace could be hosted and operated by government. A number of National Research Council (NRC) reports[24]

---

[24]NRC, 1993, *Toward a Coordinated Spatial Data Infrastructure for the Nation*, Washington, D.C., National Academies Press; NRC, 1994, *Promoting the National Spatial Data Infrastructure Through Partnerships*, Washington, D.C., National Academies Press; NRC, 1995, *A Data Foundation for the National Spatial Data Infrastructure*, Washington, D.C., National Academies Press; NRC, 2001, *National Spatial Data Infrastructure Partnership Programs: Rethinking the Focus*, Washington, D.C., National Academies Press; NRC, 2002, *Toward*

have emphasized government's role in promoting high-quality, nationwide layers of "framework" geographic data to create a basic public domain resource to meet the needs of all levels of government and the commercial sector. Interrelated initiatives such as *The National Map,* the U.S. Census Bureau's MAF/TIGER (Master Address File/Topologically Integrated Geographic Encoding and Referencing system) modernization, FEMA's floodmap modernization program, NOAA's *Digital Coast* program, Bureau of Land Management's cadastral and resource mapping programs, and Geospatial One-Stop are helping to move framework layers toward national coverage. However, the overall NSDI and *The National Map* envisage a combination of basic uniform national ("blanket") coverage *and* more patchy, varied-scale ("patchwork quilt") coverage for local data.[25] This national vision recognizes that the federal government cannot (and should not) provide more than a small percentage of the geographic data used by society.

Most visions for expanding and maintaining the NSDI stress the need to capture or access detailed local information that is already being gathered for other purposes by state, local, private, and nonprofit entities. However, past federal appeals for data donations have not always been successful. Entities are reluctant to contribute significant resources to a system that does not directly promote their own missions or needs.

A National Commons and Marketplace could provide a powerful new vehicle for soliciting donations. Sellers who use the facility could reach more buyers. This in turn would make the data vendors' existing products and services more valuable. In exchange for this service, agencies could adopt the following rule: *Creators who post a data file for sale over the "marketplace component" must at the same time deposit a copy of the data file in escrow to the secured archives of the National Commons and Marketplace. Escrowed files become available after five years through a*

New Partnerships in Remote Sensing: Government, the Private Sector, and Earth Science Research, Washington, D.C., National Academies Press; NRC, 2003, *Using Remote Sensing in State and Local Government: Information for Management and Decision Making*, Washington, D.C., National Academies Press.

[25]Blanket coverage is nationwide coverage at a uniform scale. Patchwork quilt coverage has a mixture of scales across the nation, drawing on the best available scale for a particular area. See NRC, 2003, *Weaving a National Map: Review of the USGS Vision of The National Map,* Washington, D.C., National Academies Press.

commons license selected by the creator at the time of deposit or, if no commons license is generated, enter the public domain.[26]

This "timed donation strategy" is a natural extension to current USGS policies that use licensing to draw data into the public domain.[27] Such a strategy could yield multiple benefits:

- *Offering Voluntary Participation.* Donations would be strictly voluntary. In practice, the decision to donate data would amount to a business judgment that access to the national marketplace was worth any eventual donation. This probably would happen fairly often; and because most geographic data tend to have a finite shelf life, the approach could promote the long-term expansion of the public domain and geographic information commons. No cash payment would be required, and the "data payment" would be years into the future.

- *Changing Agency Culture.* A timed donation strategy would give agencies an incentive to attract as many participants to the National Marketplace as possible. The resulting "culture shift" would foster greater sensitivity to commercial concerns and more coordination between the private and public sectors.

- *Perception of Fairness.* While often misplaced, a perception exists that it is unfair for private companies to freely and directly benefit from geographic data investments made by local, county, state, and federal agencies. Through a timed donation strategy, private

---

[26]Five years seems to be reasonable given the shelf life of most commercial products, but the definitive number should be based on a detailed study of the market.

[27]See USGS Policy 01-NMD001 (April 2001), which states that agencies should "convert licensed data to the public domain data by negotiating termination dates for license restrictions. The appropriate termination date may vary depending on the specific data type." See also the first two recommendations in National Satellite Land Remote Sensing Data Archive Advisory Committee, 2000, *Access to Restricted Data: A White Paper*, available at: <http://edc.usgs.gov/archive/nslrsda/advisory/RestrictedDataFinal.html> ("In order to fulfill its mission, the Archive may acquire restricted data as long as the restrictions expire in a specified, finite period of time" and "Accept restricted data into the Archive only with a sunset clause on every restriction; for example less than the 10-year limit exercised by Congress regarding Landsats 4 and 5 [Thematic Mapper] data. Restricted data subject to royalty arrangements should be avoided.").

companies that build on the data investments of government agencies would eventually donate their data to the public.

- *Improving Archiving.* Currently, agencies find it difficult to track data worth archiving unless the private sector notifies them that data are about to be discarded.[28] Timed donations can resolve the problem by making a copy archived with government available after a specified number of years or by making it easier to monitor data.

- *Reinvigorating the Public Domain.* A successful National Commons and Marketplace would reinvigorate the public domain by making geographic data easier to find, share, and exchange. Timed donations would increase the flow of formerly proprietary data into the public domain and commons.

Recommendation: The geographic data community should consider a system of "data donations" in which anyone who sells data using the National Marketplace in Geographic Information automatically agrees to donate their data to the commons after a commercially reasonable time, which we provisionally set at five years.

9.4.2 Private Sector as Operator

There are three generic structures that a privately operated facility could follow.

1. *Vendor-Operated Facility.* An existing vendor could operate the National Commons and Marketplace as a sideline to its core business.[29] The social value of such an enterprise would depend on its business model. Economic distortion ("deadweight loss") is smaller for models that charge users a fixed, one-time fee for using some form of the National Marketplace. For this reason, business models based on (a) one-time fees set at levels that are

[28]Agencies are, of course, aware of large commercial satellite imagery collections (e.g., SPOT data), nationwide street-centerline databases (e.g., those of Geographic Data Technologies, Inc.), and similar resources. They are much less likely to know about old digital aerial photographs or soils data.

[29]Environmental Systems Research Institute, Inc.'s *Geography Network* is suggestive of this approach.

reasonable for most that would want to participate, (b) advertising revenues, or (c) the requirement that users make reasonably priced upfront software purchases are usually benign. Inefficient markets that try to "steer" customers toward specific products, such as by favoring those products in site search algorithms, would need to be discouraged as well. Finally, a single vendor-owned National Commons and Marketplace would almost certainly raise significant antitrust issues.[30]

2. *Stand-alone For-profit Business.* The National Commons and Marketplace could operate as a stand-alone business like CommerceOne, a private corporation. A stand-alone business model would be similar to the vendor-operated case, except that a neutral third party not offering data products itself would have no incentive to "steer" consumers toward certain products. The stand-alone business model would also alleviate—but not eliminate—possible antitrust concerns.[31] Finally, a stand-alone marketplace is likely to garner consumer confidence much faster than a vendor-operator facility. This makes it potentially easier to build and reach critical mass.

3. *Nonprofit Organization.* The public policy benefits of organizing the National Commons and Marketplace as a private nonprofit organization are similar to those offered by a stand-alone business. The main difference is that nonprofit status would reduce—but not eliminate[32]—the temptation to set high access fees. A nonprofit organization also would be significantly more transparent. The public interest benefits in supporting a public commons and a timed donation strategy might be the primary basis in qualifying

---

[30]Because of "network externalities" (see Chapter 6), a successful National Commons and Marketplace would be near monopolist almost by definition. The controlling firm could potentially leverage this power to gain market dominance over data and software.

[31]A stand-alone business could still possess monopoly power. Unlike a vendor-operator, however, it would not exploit this power to dominate other markets in which it operates.

[32]Nonprofit status does not eliminate the danger of monopoly pricing. At a minimum, even nonprofit entities must break even over the long run. More fundamentally, management frequently is tempted to fund new projects and initiatives. The resulting expense tends to push most organizations toward profit-maximizing behavior. In addition, nonprofits may have incentives to divert what otherwise would be profits into perquisites for management.

> for nonprofit legal status. All of these factors would reduce the potential for antitrust violations compared to a stand-alone business.

Given current budget pressures, agencies could decide to encourage development of a privately operated National Commons and Marketplace as the best available option. If so, agencies would still need to ensure that such a private facility served the nation's geographic data needs. First, agencies would need to arrange for or accommodate virtual or physical integration between *The National Map*, Geospatial One-Stop, and related federal geographic data programs on the one hand, and the privately operated National Commons and Marketplace on the other. Second, agencies would need to ensure that a privately operated National Commons and Marketplace did not lead to significant antitrust and economic inefficiency problems. Finally, agencies would have to step in if the private sector failed to create the National Commons and Marketplace within a reasonable time frame so that government's goals and mandates were not met.

9.4.3 Division of Responsibilities

To the typical person offering data through the envisioned National Commons and Marketplace—whether intending to sell data in the marketplace or dedicate them to the commons—the license and metadata creation processes would look and feel the same. Furthermore, data searches across the commons and marketplace would be seamless—in a typical search of the virtual facility, all datasets meeting the search conditions would be returned regardless of whether licensed under a market or an open access form of license. That said, it is technically possible for different entities to host and operate different components of the system. For example, the commons component might be hosted by government while the marketplace component was hosted by a nonprofit organization. Many alternative architectures are possible. In principle, numerous commercial and government services could offer competing browsers to search for data, support transactions, and deliver data over the National Commons and Marketplace.

9.5 SUMMARY

New institutions could make licensing a more powerful and attractive tool for government agencies, commercial firms, and other affected stakeholders. The possibilities include model licenses, multiagency licenses, automated search capabilities, and an integrated National Commons and Marketplace. Because geographic data sharing and exchange relationships are complex, fundamental improvements cannot be based on a single strategy or intervention. Rather, agencies must evaluate their mandates and missions and consult constituencies to identify strategies and interventions that, taken together, yield the greatest net benefits. These benefits should extend beyond the agency and their immediate stakeholders to embrace the broader public interest.

A well-organized geographic data commons connecting users and contributors and an efficient market connecting buyers and sellers could make agency licensing more efficient, reduce wasteful duplication between agencies, accelerate the availability of local datasets in the public domain and commons, improve archiving of geographic data, increase the range of geographic data products available to consumers, and foster competition among private vendors. Such a National Commons and Marketplace might be government-operated, vendor-operated, a stand-alone for-profit business, or a nonprofit organization. Major components also could be operated by different entities. Assessment of options should address and accommodate the interrelationships and interdependencies among technical, institutional, legal, and economic issues. Whatever the chosen path, strong agency leadership will be needed to ensure maximum benefits.

Recommendation: Federal agencies should investigate options for and encourage development of a National Commons and Marketplace in Geographic Information.

10

Recommendations

1. Before entering into data acquisition negotiations, agencies should confirm the extent of data redistribution required by their mandates and missions, government information policies, needs across government, and the public interest.[1]
2. Agencies should experiment with a wide variety of data procurement methods in order to maximize the excess of benefits over costs.[2]
3. When geographic data are used to design or administer regulatory schemes or formulate policy, affect the rights and obligations of citizens, or have likely value for the broader society as indicated by a legislative or regulatory mandate, the agency should evaluate whether the data should be acquired under terms that permit unlimited public access or whether more limited access may suffice to support the agency's mandates and missions and the agency's actions in judicial or other review.[3]
4. Agencies should agree to license restrictions only when doing so is consistent with their mandates, missions, and the user groups they serve.[4]

---

[1]See Chapter 8, Section 8.3.

[2]See Chapter 8, Section 8.3.1.

[3]See Chapter 8, Section 8.3.2.3.

[4]See Chapter 8, Section 8.3.2.4.

5. Agencies that acquire data for redistribution should take affirmative steps to learn the needs and preferences of groups that are the intended beneficiaries of the data as defined by the mandates and missions of the agency. Agencies should avoid making technical choices in anticipation of secondary and tertiary uses or consumer preferences.[5]
6. Agencies should dedicate resources to training and knowledge sharing among agencies in order to extract maximum public benefit from licensing. The Federal Geographic Data Committee's working group and subcommittee structure provides a convenient venue through which agencies can report and learn from their experiences.[6]
7. Agencies, trade associations, and public interest groups should exercise leadership in promoting standard clauses and form licenses throughout the geographic data community.[7]
8. Agencies should continue to keep abreast of data brokerage and automated purchasing system developments that might help them coordinate data acquisitions from competing vendors.[8]
9. The geographic data community should consider a National Commons in Geographic Information where individuals can post and acquire commons-licensed geographic data. The proposed facility would make it easier for geographic data creators (including local to federal agencies) to document, license, and deliver their datasets to a common shared pool, and also would help the broader community to find, acquire, and use such data. Participation would be voluntary.[9]
10. The geographic data community should consider a National Marketplace in Geographic Information where individuals can offer and acquire commercial geographic data. The proposed facility would make it easier for the geographic data community to offer, find, acquire, and use existing geographic data under license. Participation would be voluntary.[10]

[5]See *id.*

[6]See Chapter 8, Section 8.5.

[7]See Chapter 9, Section 9.2.1.

[8]See Chapter 9, Section 9.2.2.2.

[9]See Chapter 9, Section 9.3.1.

[10]See Chapter 9, Section 9.3.2.

11. The geographic data community should consider a system of "data donations" in which anyone who sells data using the National Marketplace in Geographic Information automatically agrees to donate their data to the commons after a commercially reasonable time, which we provisionally set at five years.[11]
12. Federal agencies should investigate options for and encourage development of a National Commons and Marketplace in Geographic Information.[12]

---

[11] See Chapter 9, Section 9.4.1.
[12] See Chapter 9, Section 9.5

Appendixes

Appendix A

Biographical Sketches of Committee Members

Harlan J. Onsrud, *Chair*, is a professor in the Department of Spatial Information Science and Engineering at the University of Maine and chair of the Scientific Policy Committee of the National Center for Geographic Information and Analysis. He received B.S. and M.S. degrees in Civil Engineering from the University of Wisconsin and Juris Doctorate from the University of Wisconsin Law School. His research focuses on (1) analysis of legal and institutional issues affecting the creation and use of digital databases and the sharing of geographic information, (2) assessing utilization of geographic information systems (GIS) and the social impacts of the technology, and (3) developing and assessing strategies for supporting the diffusion of geographic information innovations.

Prudence S. Adler is associate executive director of the Association for Research Libraries (ARL). Her responsibilities include federal relations with a focus on information policies, intellectual property rights, telecommunications, issues relating to access to government information, and project management for the ARL GIS Literacy Project. Prior to joining ARL in 1989, Ms. Adler was Assistant Project Director, Communications and Information Technologies Program, Congressional Office of Technology Assessment, where she worked on studies relating to government information, networking, and supercomputer issues, and information technologies and education. Ms. Adler has an M.S. in Library Science, an M.A. in American History, and a B.A. in History. She has participated in several advisory councils including the Depository Library Council, the Board of Directors of the National Center for Geographic Information and Analysis, and the Advisory Committee for the National Satellite Land Remote Sensing Data Archive. From 1997 through 1999, she served on the National Research Council (NRC) Panel on Distributed Geolibraries.

Hugh N. Archer has been the commissioner for the Kentucky Department for Natural Resources since 1998. He oversees the Divisions of Energy, Conservation, and Forestry, as well as the Kentucky State Nature Preserves Commission. Previously, Dr. Archer served as Executive Director of the Kentucky River Authority (1995–1998). He has been a member of both the Kentucky and Wisconsin bar associations for over 25 years and was an attorney and planner for the Kentucky Natural Resources and Environmental Protection Cabinet for four years. Dr. Archer has consulted for government agencies, utilities, and private-sector clients on GIS issues including infrastructure management and natural resource management. He was a member of the NRC Mapping Science Committee from 1994 to 1997.

Stanley M. Besen is vice president of Charles River Associates. He was a visiting professor of Economics and Law at Georgetown University Law Center (1990–1991), a visiting Henley Professor of Law and Business at Columbia University (1988–1989), and Co-director of the FCC's Network Inquiry Special Staff (1978–1980). From 1980 to 1992, Besen was Senior Economist at The Rand Corporation, where he co-edited the *RAND Journal of Economics* (1985–1988). He currently serves on the editorial board of *Economics of Innovation and New Technology*. He has coauthored two books and authored numerous monographs, articles, and book chapters on telecommunications, media economics and regulation, and intellectual property. Dr. Besen holds a B.B.A. in Economics from the City College of New York and an M.A. and Ph.D. in Economics from Yale University. He is a member of the Committee on Internet Searching and the Domain Name System: Technical Alternatives and Policy Implications (2001–2003) and previously served on the U.S. National Committee for CODATA (1993–1996).

John W. Frazier is a professor of urban geography and co-chair of the Department of Geography, as well as director of the GIS Core Facility at Binghamton University. He has authored three books and numerous articles on the applied aspects of geography. Professor Frazier has received a number of awards, including the New York State University Professors' Service Award (1994), a national Applied Geography Specialty Group

Project Award (1994), a Kent State University Distinguished Geographer Award (1994), a Distinguished Service Award of the National Applied Geography Conferences (1995), and the James R. Anderson Medal of Applied Geography (1996), the highest honor bestowed by the Association of American Geographers for Applied Geography.

Kathleen (Kass) Green recently retired from her position as president of Space Imaging Solutions, a division of Space Imaging, LLC. While with Space Imaging, Ms. Green supervised over 200 employees, offering satellite imagery, and remote-sensing and GIS services to clients worldwide. Prior to joining Space Imaging, Ms. Green was president of Pacific Meridian Resources, a GIS consulting firm she cofounded in 1988 and sold to Space Imaging in 2000. Ms. Green's background includes 30 years of experience in natural resource policy, economics, GIS analysis, and remote sensing. She is the author of numerous articles on GIS and remote sensing and has coauthored a book on the practical aspects of accuracy assessment. Ms. Green is the recent past president of Management Association for Private Photogrammetric Surveyors, an organization of private mapping firms dedicated to advancing the mapping industry. She received a B.S. in Forestry and Resource Management from the University of California, Berkeley (1974), an M.S. in Resource Policy and Management from the University of Michigan, Ann Arbor (1981), and advanced to candidacy toward her Ph.D. in Wildland Resource Science from University of California, Berkeley. Ms. Green currently serves as an independent consultant and board member to public, private, and nonprofit natural resource and geospatial organizations.

William S. Holland is a principal, cofounder, and chief executive officer of GeoAnalytics, Inc. His expertise lies in the organizational, legal, economic, policy, and administrative aspects of GIS implementation. Mr. Holland is past president and board of directors member of the National States Geographic Information Council (NSGIC). He provided leadership to NSGIC in its development and formulation of key policy instruments underlying the development of the National Spatial Data Infrastructure. Mr. Holland has given numerous lectures and workshops at local, regional, and national conferences, and he has authored several articles about information and technology policy and intergovernmental coordination. Prior to cofounding GeoAnalytics, Mr. Holland was the first executive director of the State of Wisconsin Land Information Board. He was responsible for implementing the nation's premier program for the coordinated development of integrated geographic and land information systems across local, state, and federal agencies.

Terrence J. Keating is currently the chairman of Z/I Imaging Corporation, an Intergraph Company, and executive vice president at Intergraph Corporation, and formerly president of Lucerne International. Dr. Keating is a certified photogrammetrist, registered land surveyor, and professional engineer, and received his Ph.D. in photogrammetry and remote sensing from the University of Wisconsin, Madison. He subsequently became a professor at the University of Maine, Orono. In 1981, Dr. Keating founded Kork Systems and worked with several hundred U.S. and international mapping firms. In 1994, he sold Kork to Autometric, Inc., and served as Autometric's vice president of Commercial Products. Dr. Keating has held professional memberships and taught continuing education seminars in the American Congress on Surveying and Mapping, the Urban and Regional Information Systems Association, and the Management Association for Private Photogrammetric Surveyors. A member of the American Society of Photogrammetry and Remote Sensing (ASPRS) since 1971, Dr. Keating is currently president and has served on or chaired many ASPRS committees. He was president of the Maine Society of Land Surveyors, and his honorary society memberships include Chi Epsilon and Sigma Xi. Dr. Keating was a panelist on a National Academy of Public Administration study that developed a comprehensive, governmentwide assessment of geospatial activities, and is a former member of the NRC Mapping Science Committee.

Jeff Labonté is the director of GeoConnections Programs, a national partnership to build the Canadian Geospatial Data Infrastructure led by Natural Resources Canada (NRCan) in cooperation with federal, provincial, and territorial agencies and the private and academic sectors. He holds a B.A. in Geography and Political Sciences from Carleton University and a Masters in Public Administration. Prior to working on GeoConnections, he worked on various geospatial data integration, GIS applications, and policy projects in the Mapping Services Branch, the National Atlas, and GIS Division, all within Geomatics Canada at NRCan. He has also worked on secondment with J2 Geomatics in the Department of National Defence, developing geospatial infrastructure tools and applications.

Xavier R. Lopez is Director of Oracle's Location Services group. Dr. Lopez leads Oracle's efforts to incorporate spatial technologies across Oracle's database, application server, and eBusiness applications. He has 12 years of experience in the area of GIS and spatial databases. He holds advanced engineering and planning degrees from University of Maine, Massachusetts Institute of Technology, and the University of California, Davis. Dr. Lopez has been active in numerous academic and government research initiatives on geographic information. He is the author of a book on government spatial information policy and has authored over 50 scientific and industry publications in areas related to spatial information technology. Dr. Lopez has served on the NRC Committee on Multimodal Transportation Requirements for Spatial Information Infrastructure since 2001 and on the Committee to Review the U.S. Geological Survey Concept of the National Map since July 2002.

Stephen M. Maurer has practiced intellectual property law since 1982. He also teaches Internet law and economics at University of California, Berkeley's Goldman School of Public Policy. His research interests include academic and industry transactions, database economics, scientific data, and patent reform. Mr. Maurer's work has appeared in various journals including *Nature, Science, Human Mutation*, and *Economica*, and he has authored several articles on the protection of geospatial databases. He has spoken at conferences for organizations including the National Academy of Sciences, the National Institutes of Health, the U.S. Department of Transportation, Duke University Law School, Stanford University, the Mutation Database Initiative, and the American Association of Geographers.

Susan R. Poulter is professor of law at the S. J. Quinney College of Law, University of Utah, in Salt Lake City. She teaches in the areas of environmental law, intellectual property, and torts. Dr. Poulter holds B.S. and Ph.D. degrees in chemistry, both from the University of California, Berkeley. After a period during which she taught chemistry at the University of Utah, she received a J.D. from the University of Utah College of Law where she was executive editor of the *Utah Law Review* and was inducted into the Order of the Coif. Professor Poulter has been a member of the Council of the Section of Science and Technology Law of the American Bar Association (ABA), and has served as a section representative, ABA co-chair, and section liaison to the National Conference of Lawyers and Scientists, a joint committee of the ABA and the American Association for the Advancement of Science (AAAS). Currently, she is a member of the Advisory Board of the AAAS project on Court-appointed Scientific Experts and the Advisory Board of the AAAS project on Science and Intellectual Property in the Public Interest.

Mark E. Reichardt serves as president of Open Geospatial Consortium (OGC). He formerly was executive director of the Outreach and Community Adoption Program for OGC. Before joining the OGC, Mr. Reichardt was involved in a number of technology modernization and production programs for the U.S. Department of Defense (DoD). In the mid 1990s, he was a member of a DoD Geospatial Information Integrated Product Team (GIIPT). Under his leadership, the GIIPT validated the ability of commercial off-the-shelf hardware and software to meet many of the DoD functional requirements for geospatial production operations. In 1999, Mr. Reichardt was selected to establish and lead an international Spatial Data Infrastructure (SDI) program for the Federal Geographic Data Committee. In this position, he supported the advancement of globally compatible national and regional SDI practices in Africa, South America, Europe, and the Caribbean.

Tsering Wangyal Shawa is GIS librarian at Princeton University. In this role, Mr. Shawa is responsible for the design, launching, and management of an automated digital cartographic and geospatial information service in a campuswide networked environment. He has widespread experience in GIS data selection, software, and hardware and holds degrees in the areas of library science, geography, cartography, and education. Mr. Shawa is an active member of the American Library Association's Map and Geography Round Table (MAGERT). Currently, he is chair of MAGERT's Geographic Technologies Committee and also a MAGERT representative to the Cartographic Users Advisory Council. He was born in Tibet and has lived in several countries, including India, Nepal, Kenya, and Sudan.

National Research Council Staff

Paul M. Cutler, *study director,* is a senior program officer for the Board on Earth Sciences and Resources of the National Academies. He directs the Mapping Science Committee and ad hoc studies in the areas of earth science and geographic information science. Dr. Cutler received a B.Sc. (Hons) in Geography from Manchester University, England, an M.Sc. in Geography from the University of Toronto, and a Ph.D. in Geology from the University of Minnesota. Before joining the Academies, he was an assistant scientist and lecturer in the Department of Geology and Geophysics at the University of Wisconsin, Madison. His research is in glaciology, hydrology, and quaternary science. In addition to numerical modeling and GIS research, he has conducted field studies in Alaska, Antarctica, arctic Sweden, the Swiss Alps, Pakistan's Karakoram mountains, the midwestern United States, and the Canadian Rockies. Dr. Cutler is a member of the Geological Society of America, American Geophysical Union, Geological Society of Washington, and a fellow of the Royal Geographical Society.

Monica R. Lipscomb is a research associate for the National Academies' Board on Earth Sciences and Resources. She earned an M.S. degree in Urban and Regional Planning at Virginia Polytechnic Institute, with a concentration in environmental planning. Previously, she served as a Peace Corps volunteer in Côte d'Ivoire and has worked as a biologist at the National Cancer Institute. She holds a B.S. in Environmental and Forest Biology from the State University of New York, Syracuse.

Karen Imhof is a senior project assistant for the Board on Agriculture and Natural Resources of the National Academies. She previously worked for the Board on Earth Sciences and Resources. Earlier, she worked as a staff and administrative assistant in diverse organizations, including the Lawyers' Committee for Civil Rights Under Law, the National Wildlife Federation, and the Three Mile Island nuclear facility.

Appendix B

List of Contributors[1]

Robert Amos, City of Bakersfield, California
Ernest Baldwin, Government Printing Office, Washington, D.C.
Jonathan Band, Morrison and Foerster LLP, Washington, D.C.
Glenn Bethel, U.S. Department of Agriculture, Washington, D.C.
James Boyle, Duke University, North Carolina
Amy Budge, Earth Data Analysis Center, University of New Mexico
Michael Bullock, Intermap Technologies, Inc., Englewood, Colorado
William Burgess, Maryland Department of Natural Resources, Annapolis (*retired*)
Scott Cameron, Department of Interior, Washington, D.C.
Neal Carney, Spot Image Corporation, Chantilly, Virginia
Tom Clines, U.S. Geological Survey, Reston, Virginia
Gene Colabatistto, Space Imaging, Thornton, Colorado
Don Cooke, Geographic Data Technology, Inc., Lebanon, New Hampshire
David DeLorme, DeLorme, Yarmouth, Maine
John Faundeen, U.S. Geological Survey, Sioux Falls, South Dakota
Chris Friel, GIS Solutions Inc., St. Petersburg, Florida
Joanne Gabrynowicz, University of Mississippi School of Law
Thomas Holm, USGS-EROS Data Center, Sioux Falls, South Dakota

[1]Presentations and white papers from participants at the committee's February and May 2003 meetings are available at: <http://www7.nationalacademies.org/besr/Licensing.html>.

Rob Hudson, GIS Solutions Inc., St. Petersburg, Florida
Randy Johnson, MetroGIS, Minneapolis, Minnesota
Robert LaMacchia, U.S. Census Bureau, Suitland, Maryland
RobertaLenczowski, National Geospatial-Intelligence Agency, Reston, Virginia
Bryan Logan, EarthData Inc., Washington, D.C.
Scott McAfee, Federal Emergency Management Agency, Washington, D.C.
Patrick McGlamery, University of Connecticut Map and Geographic Information Center, Storrs.
Anne Hale Miglarese, National Oceanic and Atmospheric Administration, Coastal Research Center, Charleston, South Carolina
Charles Mondello, Pictometry, Inc., Rochester, New York
John Palatiello, Management Association for Private Photogrammetric Surveyors, Reston, Virginia
Thomas Parris, ISCIENCES, L.L.C., Boston, Massachusetts
Cindy Paulauskas, Navigation Technologies Corporation, Chicago, Illinois
James Plasker, American Society for Photogrammetry and Remote Sensing, Bethesda, Maryland
David Post, Temple University Law School, Philadelphia, Pennsylvania
William Raduchel, AOL/Time Warner (*retired*)
Jerry Reichman, Duke University School of Law, Durham, North Carolina
Barbara Ryan, U.S. Geological Survey, Reston, Virginia
Suzanne Scotchmer, University of California, Berkeley
Tim Storey, National Conference of State Legislatures, Washington, D.C.
Karl Tammaro, National Geospatial-Intelligence Agency, Reston, Virginia
Shawn Thompson, DigitalGlobe Inc., Longmont, Colorado
Mark Tuttle, State of Tennessee GIS Services, Nashville
Costis Toregas, Public Technology Incorporated, Washington, D.C.
Peter Weiss, NOAA-National Weather Service, Silver Spring, Maryland
Brian Wright, University of California, Berkeley

Appendix C

Digital Geographic Data Available in the United States

C.1 INTRODUCTION

Human curiosity about our world has generated philosophical and practical questions about nature, exploration, organization, and use of Earth's space. As a result, geographic thinking, observation and the collection of information about Earth, and mapping of Earth's features are a part of the earliest civilizations. The quest for geographic knowledge to inform our curiosity and to direct our actions is reflected in maps that address pressing spatial questions. Such questions have become more complex during the past century due to discovery, technology, and human treatment of the physical environment, including the rapid industrialization and concomitant urbanization of the globe. They also have taken on a more pressing drive for fast answers. Fortunately, our ability to acquire, store, retrieve, visualize, and analyze large geographic databases, containing location-specific identifiers that link features to geography, also has increased dramatically. In fact, it might be argued that technological improvements fostered data generation efforts by the end of the twentieth century. Today, digital devices enable the rapid acquisition and maintenance of an incredible range of geographic databases, a vast inventory of information about Earth, ranging from geodemographic descriptors to well-defined uses of small plots of land. Barely recognized by local planners a quarter century ago, geographic information systems (GIS) are now in general use across the United States.

As we enter a discussion of geographic data licensing, it is necessary to be aware of the types of data now available. The purpose of this chapter is to classify geographic data to permit discussion of a broad range of geographic data types now available in the United States from a variety of sources, public and private. We broadly classify geographic data types by their origin, resulting from either processes of the natural world or human action. As any taxonomy, ours has fuzzy parts. We know that the physical and human worlds are intertwined and that human actions influence physical process and patterns. Nonetheless, a simple taxonomy allows for easy discussion of the various types of geographic data. We also associate federal agencies with particular types of geographic data, recognizing that agencies with parallel or overlapping functions at other government levels also use such data types.

C.2 GEOGRAPHIC DATA AND THE PHYSICAL WORLD

Numerous features and variables are used to describe, visualize, analyze, and monitor Earth's physical processes, patterns, and conditions. Of growing importance is the use of geographic data. In the United States, numerous government agencies and commercial firms participate in the acquisition and applications of geographic data. Below, we summarize features related to weather, hydrology, elevation, geology and physical geography, energy, and hazards that are available as geographic data.

C.2.1 Weather/Meteorological Data

Meteorological processes and weather influence biological and ecological systems, including the growth and health of species. Weather and climate also affect long-term geological and geomorphic processes and the physical geography of Earth's surface. Table C-1 reports four weather features—precipitation, temperature, humidity, and winds—that often are presented as geographic data and used to assess current conditions and predict future ones. Federal agencies acquire and apply these geographic data. In addition, government agencies contract for commercial services related to data acquisition, and so, the private sector also is a stakeholder in these processes.

TABLE C-1 A Classification of Geographic Data Available in the United States: The Physical World

Topic	Features	Examples of Federal Stakeholders[a]
Weather	Precipitation, temperature, winds, and humidity	DOE, FEMA, NOAA
Hydrology	Water bodies, drainage/subsurface, wetlands, and watershed/drainage basins	EPA, FWS, NRCS, USDA, USGS
Elevation/bathymetry	Slope, contours, aspect, and digital elevation models (DEMs)	BLM, NOAA, NRCS, USDA, USGS
Physical geology and physical geography	Bedrock geology, surficial geology and geography, soils, and land cover	BLM, DOE, NRCS, USDA, USGS
Energy resources	Coal deposits, oil fields, natural gas reservoirs, geothermal fields, other natural resource deposits	BLM, DOE, EIA, NRCS, USGS

[a]BLM = Bureau of Land Management, DOE = Department of Energy, EIA = Energy Information Administration, EPA = Environmental Protection Agency, FEMA = Federal Emergency Management Agency, FWS = Fish and Wildlife Service, NOAA = National Oceanic and Atmospheric Administration, NRCS = Natural Resources Conservation Service, USDA = U.S. Department of Agriculture, USGS = U.S. Geological Survey.

C.2.2 Hydrology

A large number of environmental surface features related to water and its movements occur as geographic data. At least six federal agencies are stakeholders in their acquisition and use. Surface water bodies in the United States include rivers, streams, lakes, ponds, reservoirs, and human-made canals. The subsurface (drainage) includes all areas into which water drains and water in various locations below Earth's surface, such as wells or springs that emit water and aquifers (underground beds).

An additional water-related category includes lowland areas saturated by water (e.g., marsh, swamp) and considered wildlife natural habitats designated as wetlands. Finally, watersheds/drainage basins are areas draining to a common waterway, such as a stream or lake (see Table C-1).

In addition to the natural features, monitoring, planning, and intervention activities require specially designated geographic regions, or districts, that are reported as geographic data (see Table C-1). Geographic data facilities devoted to measuring water composition, flow rates, and depth include water-monitoring facilities, dams, gauging stations, and rainfall/precipitation stations. Geographic regions, specified by local, state, or federal actions, including water districts, floodplains, and historic flood zones and points that have recorded 100-year and 500-year floods, fall into this category.

C.2.3 Elevation and Bathymetry

Elevation and bathymetry are the heights above or below sea level, respectively. Related features include slope, contours, and DEMs. Natural landscapes contain sloping surfaces, implying curvature of surface as one moves from location to location. Slope is a measure of angular change between elevation points on a continuous surface. Contours are lines on a map that connect points of equal elevation. A DEM is a database of point measurements of elevation at regular intervals. The most common forms include feature points separated by 10 meters, although small-scale maps use larger intervals. Contour line and slope surfaces typically are derived from DEMs. All three features associated with elevation can be imported into most GIS.

C.2.4 Physical Geology and Physical Geography

The natural environment includes surface and subsurface features that result from long-term Earth processes and are important for understanding the effects of human actions. Together, these features constitute the physical geology and physical geography of the natural environment. Included in such features are bedrock geology, surficial geography and geology, and soils and land cover (see Table C-1).

Bedrock geology, including features that are located beneath unconsolidated, depositional matter and soils, may be important for engineering and construction projects. Surficial features refer to various landforms and depositional characteristics of Earth's surface. Soils are classified by their physical characteristics and vary widely. Despite limited sampling to verify the generalized areas of similar soil types, digital soils maps are widely used for planning and development applications.

C.2.5 Energy Resources

Energy resources hold substantial locational importance. Among the geographic data available for this category are coal deposits, oil fields, geothermal fields, and natural gas reservoirs. The latter include a number of gases typically found in association with petroleum deposits, including methane, ethane, propane, and butane.

C.2.6 Natural Hazards

Naturally occurring events can become hazardous when human settlements coincide with their locations. These natural hazards include tornadoes, hurricanes, floods, volcanoes, and earthquakes.

C.3 GEOGRAPHIC DATA AND THE BUILT ENVIRONMENT

The built environment constitutes the human geography at Earth's surface. Both visible and invisible features are important ways of organizing living space. These include overt structures that define cultural landscapes and invisible boundaries that reflect politic, economic, and locational decisions. These data are broadly summarized into eight major categories (Table C-2): transportation-related, institutional locations, energy-related infrastructure, administrative and legal, hazardous locations, business, communications, and health.

C.3.1 Transportation

Geographic transportation data are related to infrastructure, routing, roadway descriptions, and special projects. Transportation technology was crucial to the early growth of the U.S. economy and metropolitan system. It continues to be a crucial part of a successful economy. The U.S. transportation infrastructure is complex and contains both national and regional linkages and features that are points and areas distributed within and around its infrastructure. The linkages contain major and minor roads and highways, including the U.S. Interstate Highway System, and a national network of railways that link major cities and resource areas with production areas. Other transportation linkages include the intracoastal waterways—the canal and river systems that serve as commercial linkages—bridges that connect any of these linkages, and local and regional transit arteries, such as subways.

TABLE C-2 A Classification of Geographic Data: The Built Environment

Topic	Features	Examples of Federal and Other Stakeholders[a]
Transportation	Infrastructure, routing, roadway descriptions, and special projects	ACE, BTS, CENSUS, DOT, EPA, FAA, FWS, FHA, FTA, FRA, NRCS, NHTSA, NOAA, NASA, USDA
Energy generation sources and transmission	*Generation*: hydroelectric facilities, oil rigs, wind farms, nuclear power plants	ACE, BLM, DOE, EIA, FEMA, USGS
	Transmission: transformers and transfer stations and power transmission lines	DOE, EIA
Institutional	Colleges/universities, schools, libraries, churches, hospitals, nursing homes, parks, industrial sites, historical sites/districts	BLM, Census, EPA, USGS
Administrative and legal	Legal, legislative, census, geography and special-purpose boundaries, cadastral	Census, HUD, Commerce

Hazardous sites and hazardous materials	Test and monitoring sites, landfills, hazardous waste sites, toxic release inventories	ACE, BLM, DOE, EPA, EIA, NRCS, USGS
Business	Geodemographic data, industrial locations, retail/wholesale businesses	Census, ESRI, MAPINFO, etc.
Communications	Telegeography, TV stations, fiber optics, etc.	PriMetrica, Inc., Equinix
Health	Health indicators, disease incidence, low birthweight, etc.	CDC, state departments of health

[a]ACE = Army Corps of Engineers, BLM = Bureau of Land Management, BTS = Bureau of Transportation Statistics, CDC = Centers for Disease Control and Prevention, Census = U.S. Census Bureau, Commerce = Department of Commerce, DOE = Department of Energy, DOT = Department of Transportation, EIA = Energy Information Administration, EPA = Environmental Protection Agency, ESRI = Environmental Systems Research Institute, Inc., FAA = Federal Aviation Administration, FEMA = Federal Emergency Management Agency, FHA = Federal Highway Administration, FTA = Federal Transit Administration, FRA = Federal Railroad Administration, FWS = Fish and Wildlife Service, HUD = Department of Housing and Urban Development, NASA = National Aeronautics and Space Administration, NHTSA = National Highway Traffic Safety Administration, NOAA = National Oceanic and Atmospheric Administration, NRCS = Natural Resources Conservation Service, USDA = U.S. Department of Agriculture, USGS = U.S. Geological Survey.

In addition to these major transportation arteries, other transportation points and areas that constitute important components of the overall infrastructure include airports, bus stations, railroad stations, highway exits and toll plazas, service and rest areas along highways, and gasoline and fueling stations. Finally, port facilities constitute an important part of the U.S. transportation infrastructure and thus are important elements of geographic data.

A significant part of transportation is the repetitive vehicular travel along paths and corridors. Such routes apply to public and private carriers, and schedules on such routes are well established. Bus, truck, rail, ferry, and airline routes are examples of features that are captured as geographic data.

Some geographic data features depict important aspects of roadways. Line feature data depicting the approximate center of a road is an important example and is a data element built into the U.S. Census Bureau's Topologically Integrated Geographic Encoding and Referencing system (TIGER) files. Finally, special transportation projects are sometimes captured as geographic data. An example is a construction project undertaken by DOT at a particular location.

Geographic transportation data are extremely comprehensive, and therefore, it should come as no surprise that more than a dozen federal agencies (Table C-2) are stakeholders in the acquisition and use of these data.

C.3.2 Energy Generation and Transmission

Power sources and their transmission are vital to the national economy. In this category, we include human-made facilities and the means of their transmission to intermediate points and to end users. Examples of human-made energy generation locations include hydroelectric and nuclear power plants, oil rigs, and wind farms. Examples of geographic data describing features of energy transmission include transformers, transfer stations, and transmission lines. Such data play multiple roles in calculation efficiencies, creating cost-saving alternatives, and monitoring systems.

C.3.3 Institutional

Institutions play vital roles in perpetuating culture; educating the population; and providing public and private services, places to worship, and historic preservation. As such, this is a broad category containing numerous cultural features that are part of the built environment that are of significance because they provide continuity for the long term, shaping cultural beliefs and traditions, and facilitate some daily behaviors of Americans.

Examples of such institutions appear in geographic databases, including (1) places of learning such as colleges/universities, schools, and libraries; (2) medical and extended-care facilities such as hospitals and nursing homes; and (3) other locations, including industrial sites, parks, and historic landmarks and sites (see Table C-2). In the latter case, a National Register exists as part of a federal program to coordinate the identification, evaluation, and protection of American historic and archeological resources.

C.3.4 Legal and Administrative

The U.S. government has created or recognized legal boundaries that provide the framework for a host of government actions and individual behaviors. These legal and administrative boundaries are political boundaries in that they establish the area in which certain laws and responsibilities may take place under the jurisdiction of elected or duly appointed officials. These geographic data include state, county, city, town, and place definitions that have legally or administratively recognized boundaries.

Within the legal boundaries are a set of legislative boundaries that provide the framework for representative government. These include election districts, congressional districts, assembly districts, and senate districts. These not only form the geographic means of representation; they also are utilized for voting trend analysis and postelection assessments.

U.S. Census Bureau TIGER files contain another form of administrative boundary that is widely used in census geography. These files contain the previously discussed legal and legislative boundaries, such as local school districts, and also contain a hierarchy of urban geographic zones, including metropolitan statistical areas, census tracts, block groups, and blocks suitable for a wide range of analyses. These boundaries often are utilized to create special-purpose areas/districts and supporting geographic databases. Examples include low- to moderate-income designations in urban centers, economic development and enterprise zones, and agricultural districts within counties.

There are many other types of special-purpose districts created from census or cadastral maps. An example is the local zoning map, which creates geographic areas of permitted land-use activities and limits other types of land use and development elsewhere.

Cadastral refers to geographic units of land that have been legally defined by professional survey standards for the purpose of land ownership. Thus, a great deal of geographic data has been accumulated on a local basis for the purposes of physical description and local taxation. Although the volume of data collected and maintained may vary, most systems contain land dimensions and area, number and type of structures, value of the land, and value of the structure. Of all the data described thus far, these are the only data collected, managed, and controlled by local administrative and legal rules.

C.3.5 Hazardous Sites

Environmental awareness and governmental actions have resulted in the collection, maintenance, and analysis of data on hazardous sites and their features. For example, EPA maintains a network of monitoring sites where water and air samples are collected routinely for analysis. The locations of nuclear power sites are also maintained as geographic data.

Environmental issues related to the disposal and storage of human waste have also resulted in geographic databases. For example, since harmful chemicals can seep into groundwater where refuse is stored, landfills are reported as geographic data. Similarly, the use of chemicals for fertilizers, sites of toxic releases and accidents, the use of herbicides, and the deposition of manure are all included in geographic database inventories maintained by USGS, EPA, and other agencies.

C.3.6 Business Geographic Data

Businesses increasingly use geographic data to target and evaluate markets, conduct site location analyses, and assess store performance. Geodemographic data are essential to these enterprises and represent census geography and special market areas. In addition to demographics, the location of competing retailers and affiliated businesses are typically geocoded geographic databases available from private data providers used in various types of business analyses.

C.3.7 Communications and Geographic Data

The locations of TV stations, fiber-optic lines, and forms of telegeography have become available in recent years from private vendors and are widely used in the communications industry. Vendors include Primetrica, Inc., and Equinix.

C.3.8 Geographic Health Data

The geographic distribution of disease and the use of health care indicators in geographic analyses are two examples of geographic health data. The availability of health indicators, such as low birthweight, in association with toxic waste storage, has led to equity studies of environmental health issues. Also, the distribution of disease and health conditions, such as HIV/AIDS, are important examples of geographic health databases.

C.4 IMAGES AS GEOGRAPHIC DATA

Geographic data capture and use are based on two models: raster and vector. Vector data represent real-world themes such as addresses (points), road networks (lines), or land parcels (polygons). The location and shape of the polygon, line, or point is determined by the coordinate position (e.g., latitude/longitude) of its node(s). Many of the data types in Tables C-1 and C-2 are examples of vector data. Raster data organize a geographic space into cells, with each cell containing attribute measures (e.g., reflectivity, elevation, frequency, concentration). Each cell in the grid is of equal size and the cell density establishes the *spatial resolution* of the grid (see Figure C-1). This class of geographic data includes a large range of layers derived from digital aerial photographs, x rays, sonar, and thermal images remotely detected by sensors on aircraft or satellites. Commercial images vary widely in resolution, cost, and use. A high-resolution aerial photo can provide a resolution whereby each pixel represents a square on the ground of only a few centimeters on a side, whereas the highest resolution commercial satellite images have pixels that represent squares measuring 0.62 meter on a side.

Images have a number of uses related to geographic data. Sometimes, for example, they provide the basis for "heads-up digitizing" of planimetric features (building footprints, streets, etc.). Images also are frequently used as "backdrop" for the display of other geographic data. Images also can be utilized for the derivation of land-use and land-cover layers for multiple uses.

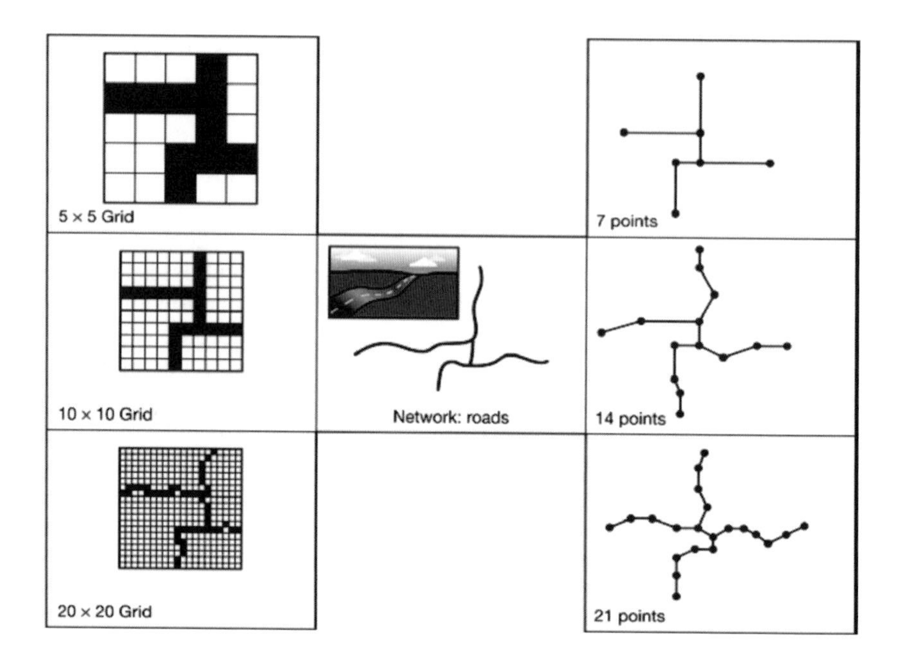

FIGURE C-1 A road network represented at three spatial resolutions by raster (left) and vector (right) data. SOURCE: I. Heywood, S. Cornelius, and S. Carver, 2002, *An Introduction to Geographical Information Systems,* 2nd Ed., Upper Saddle River, N.J., Pearson Education, p. 53, ©2002. Reproduced with permission of Pearson Education, Inc., Upper Saddle River, N.J.

APPENDIX D

Selected Licensing Models
Selected Licensing Alternatives
and
Clearview Contract

D.1 Selected Licensing Models

Type of Restriction	Description	Remarks
A. Restricted Users Lists		
Subcontractor	Data limited to users working for licensee plus subcontractors, agents, and/or consultants.	Common term when government acquires data under license. Some licenses require government to notify vendor before transferring data to subcontractor.
Predefined user lists	Data limited to users working for licensee plus specified entities.	Space Imaging Multiagency License allows government to purchase 1-, 5-, and 10-agency licenses.
		National Geospatial-Intelligence Agency's (NGA's) Clearview license permits agency to redistribute data to state, local, and foreign governments and nongovernmental organizations engaged in joint research projects. Recipients may not use images for their own purposes and are required to return data after intended use is completed.
Predefined categories	Data may be redistributed to predefined categories of users.	Maryland's Model Procurement License allows agencies to redistribute data to academic and educational institutions.

Uplift fees	User may redistribute data to specified classes of individual in return for pre-agreed fee.	Common term when government acquires satellite data. U.S. Geological Survey (USGS) Draft Model Contract (Purchase of Satellite Data) would allow USGS to upgrade its license to include a wider class of users by paying a pre-agreed fee.
Best efforts	User undertakes to limit redistribution to maximum extent possible.	NGA Clearview license obligates agency to "minimize the sharing of imagery with entities who would otherwise purchase the imagery."
B. Restricted Uses		
Specified project(s)	Data may only be used for one or more enumerated projects.	Maryland State Geographic Information System (GIS) License allows agencies to publish/present graphics and tabular material derived from vendor data. Space Imaging Landsat license lets users reproduce imagery in journals provided that proper acknowledgment is given. National Oceanic and Atmospheric Administration (NOAA)/Space Imaging Ikonos license allows agency to distribute images in "a non-manipulable

		(e.g. bitmap) format as part of a research report or publication."
Enumerated uses	Data may only be used for specific applications.	NOAA Sea-viewing Wide Field-of-view Sensor (SeaWiFS) data restrict agency to use data for "civil marine operational purposes." County government licenses to Federal Emergency Management Agency (FEMA) require agency to use data for developing Flood Insurance Map.
Disaster response	Data may only be used for disaster response.	FEMA routinely licenses the right to distribute data "for a limited period" following disasters.
Education	Data may only be used for education.	USGS Policy 10-NMD001 (April 2001): "Whenever possible, agreements should allow the unrestricted use of . . . data for disaster response, research, or educational purposes.
Research	Data may only be used for research.	USGS Policy 10-NMD001 (April 2001): "Whenever possible, agreements should allow the unrestricted use of . . . data for disaster response, research, or educational purposes.
Government uses	Data may only be used for government purposes.	Traditional National Imagery and Mapping Agency (NIMA) satellite licenses gave agency the right to release reduced-scale images for any federal government purpose.

Noncommercial uses	Data may only be used for noncommercial purposes.	NOAA/Space Imaging license permits agency to redistribute data "on an isolated, non-commercial basis." NGA Clearview license permits government to release a "limited number" of hardcopy scenes for "public information, diplomacy, emergencies, disasters, and other non-commercial uses."
C. Restricted Locations		
Machine	Data may only be stored and accessed on specific computers and/or terminals. Licensee is frequently permitted to make a single backup copy.	—
Site	Data may only be stored and accessed at one geographic location.	Traditional method for restricting redissemination. U.S. Government Printing Office argues that "At a minimum, users should be able to access and download data for re-use, at no charge, in a federal depository library." Some state legislatures put redistricting data in special kiosks so that the public can access them.

Entity	Data may only be stored and accessed at geographic locations belonging to a specified entity.	NGA Clearview license extends to all federal agencies and other specified partners. USGS Systeme Probatoire Pour l'Observation de la Terre license contains similar provision.
D. Derivative Product Restrictions		
Reformatting	Users cannot convert vendor data to unsupplied formats.	—
Products for internal use	Users can produce unlimited derivative products but cannot redistribute them outside licensee's organization.	NOAA/Space Imaging license.
Specified products	Users can extract data to make enumerated derivative products.	NOAA/Space Imaging license.
Quantity limits	Users can utilize limited amounts of data to make derivative products.	Maryland Model Procurement License allows agencies to share data covering areas less than or equal to a single county.
Noninvertible products	Users can make derivative products that cannot be inverted.	Standard satellite license term. USGS Draft Model Contract (Purchase of Satellite Data) provides that government can freely

		redistribute any derived product provided that "the original source image data or a reasonable facsimile thereof is not included as a file, layer, component, or other accessible and/or viewable portion of the derived product, whether this data is separate or combined with other data and/or information."
Value-added products	Licensee can use data to create derivative products that require substantial additional effort to create.	Canada's Radarsat license allows licensee to produce value-added products that exclude significant additional interpretation and/or data. "Value-added products" do not include mosaics, geocoding, subscenes, and various other specifically enumerated exceptions.
E. Redistribution Restrictions		
Reduced-resolution data	Users can distribute data at reduced resolution.	Satellite licenses commonly let government agencies post reduced-resolution data on the Internet. USGS Policy 01-NMD001 (April 2001): Government should "pre-negotiate defined USGS-derived products that can be generated from licensed data and placed in public domain. In a mapping context, derived products could include vectors digitized from licensed imagery, DEMs produced by thinning or generalized licensed DEM

		source, or degraded imagery produced from full resolution licensed source."
Embargoed data	Users can distribute data after fixed period of time.	National Aeronautics and Space Administration (NASA)/Orbital Image Corp. agreement grants company exclusive right to use and sell SeaWiFS ocean color data for two weeks. Thereafter, NASA may redistribute data to scientists.
		USGS Policy 01-NMD001 (April 2001): Agency may negotiate licenses that "[c]onvert licensed data to public domain data by negotiating termination dates for license restrictions. The appropriate termination date may vary depending on the specific data type."
Selected attributes data	Users can extract enumerated attributes for use in building derivative products.	U.S. Census Bureau/ Geographic Data Technologies, Inc. (GDT) license permits agency to extract selected features and attributes from company's DYNAMAP product to support public domain Topologically Integrated Geographic Encoding and Referencing system database.
		USGS Policy 01-NMD001 (April 2001): Government should "pre-negotiate defined USGS-derived products that can be generated from licensed data and placed in public domain. In a mapping context, derived products could include vectors

		digitized from licensed imagery, DEMs produced by thinning or generalized licensed DEM source, or degraded imagery produced from full resolution licensed source."
Nonmanipulable data	Users can distribute data in formats (e.g., paper, PDF) that cannot be manipulated.	NOAA/Space Imaging license permits agency to distribute images in a "nonmanipulable (e.g. bitmap) format as part of a research report or publication."
Display-only	Users can distribute data in nondownloadable form over the Internet.	USGS Draft Model Contract (Purchase of Satellite Data): Agency may post source data and derived products over the Internet in formats that prevent downloading or screen capture.
		Traditional NIMA License: Agency may post images on Internet "provided implementation precludes downloading or screen capture."
Logos and notices	Users can distribute data provided that vendor's identity and/or license restrictions are included.	Maryland Model Procurement License: Licensee must include notice reciting license restrictions whenever product is redistributed.
		NOAA/Space Imaging license: All licensed images must bear vendor's copyright notice.
		USGS Draft Model Contract (Purchase of Satellite Data): Agency may display images and distribute

		derivative products containing vendor's data provided that "conspicuous copyright and license notices appear."
Data buys	Licensee receives unlimited right to use and redistribute data.	Functionally equivalent to nonlicense purchase agreement.
		Maryland Department of Natural Resources/PIXXURES license. Agency receives perpetual unlimited license to use and redistribute orthophotography images, including Internet posting rights.
F. Nonroyalty Terms		
Integrity	Users cannot manipulate and redistribute data.	Limits potential liability for postpublication alteration of data.
Attribution	Users who redistribute data must give proper credit to original vendor.	Facilitates agencies' ability to document value of government datasets. Enhances private-sector vendors' ability to build name recognition and reputation.
		Maryland Model Procurement License: Licensee must include notice reciting license restrictions whenever product is redistributed.

		Space Imaging Landsat license lets users reproduce imagery in journals provided that proper acknowledgment is given.
Risk management	Users agree to assume liability through liability, indemnity, and/or attorneys fee agreements.	Facilitates agencies' financial ability to distribute data at zero cost.

D.2 Selected Licensing Alternatives

Business Model	Description	Remarks
Consulting/fee-for-service	Vendor sells unlimited right to use and redistribute data.	Traditional model. Still used by most aerial survey firms. Vendor typically retains nonexclusive right to resell data to third parties. Resale may be economically important where would-be buyers have no practical option to identify original purchaser.
Update services	Vendor improves quality and/or timeliness of data at regular intervals.	Many customers place a substantial premium on small quality improvements and/or regular updates.
Branding	Vendor's reputation for reliability commands a premium from consumers.	
Advertising and content	Vendor's data draws users to Web site and/or is linked to advertising or other Web sites.	Yahoo offers "free" geographic data to attract users to its site.
Copyright only	Vendor relies on copyright protection only.	Traditional distribution model for paper maps.

		California state agencies routinely copyright seismic fault-line databases to prevent unauthorized revisions.
Bundled content	Vendor bundles unprotected data with copyrighted software tools	Many companies bundle software with government data. Examples include Caliper, Warren Glimpse, and DeLorme. Maptech Raster Nautical Charts cannot be viewed without the company's copyrighted software.
Distribution services	Vendor delivers public domain data in easy-to-locate, convenient package.	Wessex and Warren Glimpse are examples of vendors who built successful businesses by redistributing agency data at lower cost and/or in more convenient formats.
Market-Maker		
Data aggregator	Assembles permissions to existing datasets in order to make and sell new products.	GDT offers clients nationwide support assembled from data trades with local and county governments.
Index services	Vendor provides convenient, one-stop access to large amounts of data.	TerraServer and Ikonos offer low-resolution image libraries to consumers free of charge. Consumers can purchase more detailed images for a fee.

		Environmental Systems Research Institute, Inc. hosts an online market for data products. Consumers who make purchases pay a transaction fee. USGS has discussed a similar scheme in which it would collect download fees for vendors who post data on *The National Map*.
Transactions support	Consultants assemble permissions needed to make new products by combining and modifying preexisting data sets.	Harris Corp. obtains street-centerline data for U.S. Census Bureau by, inter alia, obtaining permissions from local and county governments.
Technical Protections		
Encryption	Data are encrypted to restrict distribution to particular computers and/or to prevent unauthorized manipulation and extraction.	California state agencies routinely encrypt seismic fault maps to prevent unauthorized revisions.
Watermarking	Data contain hidden "steganographic" data that identify vendor.	Allows licensor to trace and prove misappropriation of data. Maryland Model Procurement License: Agency may insert "tracking and authentication devices" to ensure contract compliance. Licensees agree not to remove them.

Password protection	Data can only be accessed by known users.	—
Download limitations	Users can only obtain small portions of total dataset.	—

D.3 CLEARVIEW CONTRACT[1]

D.3.1 Background

In January 2003, NIMA (now NGA) signed a nonexclusive licensing agreement with U.S. satellite companies Digital Globe, Space Imaging, and ORBIMAGE to procure high-resolution imagery. The contract has a $500 million ceiling for each company over its five-year life span. Using NGA's bargaining power, Clearview negotiators aimed to replace multiple government licenses with a single license, and promote stability in the U.S. commercial satellite industry.

D.3.2 Types of Imagery

Clearview covers Panchromatic (black and white), multispectral (color), and "other remotely sensed data." The contract also contains options for value-added imagery processing, external purchases, and direct downlink purchases. Space Imaging's Ikonos, launched in 1999, and ORBIMAGE's OrbView3, launched in 2003, have 1-meter panchromatic and 4-meter multispectral options.[2] DigitalGlobe's Quickbird camera captures 0.61-meter panchromatic and 2.44-meter multispectral imagery.

D.3.3 Who Can Use the Data?

Clearview affords unrestricted access to the data by the U.S. government (all branches, departments, agencies, offices, and contractors therewith). Additionally, state and local governments, foreign governments, intergovernmental organizations, nongovernmental organizations, and other nonprofit organizations have unrestricted access when working with the U.S. government on "joint projects." Such projects are defined as coalition force operations, relief efforts, homeland security operations, exercises, and co-production. Activities including city planning, property

[1]This section is based on testimony of Karl Tammaro, NGA.
[2]See <http://www.spaceimaging.com>; <http://www.orbimage.com/news/releases/06-26-03.html>.

tax assessment, transportation infrastructure management, and "general purpose mapping" are not considered "joint projects," and are excluded.

D.3.4 Distribution Restrictions

Imagery cannot be placed on an electronic distribution system that permits access by unlicensed users. Additionally, derived products containing imagery data inherit the copyright and license restrictions of the source data.

D.3.5 Public Availability

Reduced-resolution data with 16-meter ground resolution or coarser retain copyright markings, but have no restrictions on use or distribution. During emergencies, disasters, or for diplomacy or public information, a "limited" number of hardcopy imagery scenes or softcopy samples may be released by a licensed user. However, commercial uses, resale, or mass public distribution are not permitted. Hard and soft copies of imagery (with the copyright mark) may be shown but not given to unlicensed users.

D.3.6 Data Archiving

Partners in joint projects with the U.S. government cannot retain the data after completion of the project. The data are archived at NGA.

D.3.7 Benefits

Industry has a five-year contract with minimum guarantees ($120 million to Space Imaging and $72 million to Digital Globe) over the first 3 years, and two 1-year renewal options. The U.S. government acquires the data at lower cost,[3] and fewer resources were expended on contract negotiations by both sides, when compared with negotiating multiple licenses.

---

[3]The bulk purchase afforded a 75 percent price break (Gene Colabatistto, Space Imaging, personal communication, December 2003).

D.3.8 Limitations

Public access to original imagery is prohibited. Partners on joint projects may not retain data for non-security-related business operations; if the data are needed, they must be purchased under a separate license.

D.3.9 Civil Agencies and Commercial Satellite Companies

The 2003 White House Directive on Commercial Remote-sensing Policy instructs civil agencies to first consider U.S. satellite companies when weighing options for imagery purchases. Clearview or a "Clearview-like" contract is being advocated by industry (Gil Klinger, speaking at NASA headquarters on June 26, 2003) to simplify their contract negotiations with civil agencies. Unlike the military sector, however, the civil sector has no single mapping agency through which to focus purchasing. At the time of writing, discussions were being coordinated among USGS, NASA, and NOAA, and led by USGS.

D.3.10 Nextview

A $500 million award was made by NGA to DigitalGlobe in October 2003 under the "Nextview" contract.[4]

[4]See <http://www.nima.mil/ast/fm/acq/093003.pdf>.

Appendix E

Glossary

"BEST EFFORTS" CLAUSE. A clause that says that the licensee will use its "best effort" to exercise its contract rights in ways that preserve the licensor's ability to earn revenues from additional licenses or sales.

BRIGHT-LINE RESTRICTION. A restriction in a contract that is easy and unambiguous to apply. Measurable quantities—7.2 meters, 1000 Angstroms—are an example.

BUSINESS-TO-GOVERNMENT PURCHASING SYSTEMS. Systems that enable automated purchasing of standardized commercial products by government.

CADASTRAL DATA. Data that describe the rights and interests in property.

CLICK-WRAP LICENSE (see also SHRINK-WRAP LICENSE). A license setting forth the terms under which a vendor sells a right to use a product; the license typically accompanies the electronic file containing the licensed data, or is on the vendor's Web site.

COPYRIGHT. Exclusive legal right to reproduce, publish, and sell works such as databases, datasets, maps, images and other works that incorporate creative expression, and software; copyright will not protect individual facts or even compilations of facts that do not have an original selection and arrangement

DATA. Facts and other raw material that may be processed into useful information.

DATA BROKERAGES. Institutions that enable users to search for previously licensed data.

DIGITAL LINE GRAPH. Line-map information in digital form.

FOUNDATION DATA. Geographic data at the foundation of government business: terrain (elevation) data, orthoimagery, and geodetic control (see separate definitions of the last two).

FRAMEWORK DATA. Frequently used data in many government applications, including transportation networks; political, administrative, and census boundaries; hydrology; cadastral data; and natural resources data.

GEODETIC CONTROL. Common reference system for establishing coordinate positions (e.g., latitude, longitude, elevation) for geographic data.

GEOGRAPHIC DATA. Any location-based data or facts that result from observation or measurement, or are acquired by standard mechanical, electronic, optical, or other sensors.

GEOGRAPHIC INFORMATION. Geographic data or works without distinction, which may encompass, but is not limited to (1) location-based measurements and observations obtained through human cognition or through such technologies as satellite remote sensing, aerial photography, Global Positioning System, and mobile technologies; and (2) location-based information transformed as images, photographs, maps, models, and other visualizations. Geographic data and works are not strictly location-based but also may include, for example, spatial relationships, descriptions or attributes of geographic features, metadata, and additional types of information that are arranged, categorized, or accessed in reference to their geographic or spatial location. Such information is typically in digital form and may be contained in databases.

GEOGRAPHIC INFORMATION COMMONS. A system for making geographic data and works openly and freely accessible to the public over the Internet. A geographic information commons may include both public domain (i.e., free from any use restrictions) and open access content (i.e., content openly available for others to access, use, and copy, and often to make derivative works although some limited restrictions may apply).

GEOGRAPHIC INFORMATION MARKETPLACE. A system for making geographic data and works available for sale over the Internet.

GEOGRAPHIC SERVICES. Processes of obtaining, processing, or providing geographic data or geographic works. As used in this volume, the term refers to the provision of access to and use of preexisting data or databases, such as subscription to a particular online geo-based processing capability or subscription to a database allowing downloads when desired. In some contexts, the term "services" may connote geographic data or works provided for a single client, according to that client's specifications.

GEOGRAPHIC WORKS. Works incorporating geographic data that have been collected, aggregated, manipulated, or transformed in some manner. Examples include datasets and databases, and other products derived from geographic data, including but not limited to maps, models, and other visualizations involving geographic data.

HYDROLOGY DATA. Location, geometry, and flow characteristics of rivers, lakes, and other surface waters.

INFORMATION COMMONS. Public domain content (free from any legal rights protection, e.g., ideas, publicly known facts, and intellectual works in which copyright has expired) and open access content (openly available to anyone but some use conditions are controlled by license).

LICENSE or LICENSING *(as used in this report).* A transaction or arrangement (usually a contract, in which there is an exchange of value) in which the acquiring party (i.e., the licensee) obtains information with restrictions on the licensee's rights to use or transfer geographic information.

MANDATE. A required function that is defined by law, typically statute, administrative code, or case law.

MARGINAL COST. The cost of providing a copy to an additional user.

METADATA. Information about data; for example, it might record such details as the collector, the sensor used, and when the data were collected (see Federal Geographic Data Committee, 1998, *Data Content Standard for Digital Geospatial Metadata*, available at <http://www.fgdc.gov/standards/documents/standards/metadata/v2_0698.pdf>).

MISSION. Either a discretionary function or an approach to accomplishing a mandated function that is carried out as part of a strategic or operational direction.

NATIONAL SPATIAL DATA INFRASTRUCTURE. Technologies, policies, and people necessary to promote sharing of geographic data throughout all levels of government, the private and nonprofit sectors, and the academic community.[1]

OLIGOPOLY. A form of imperfect competition in which there are relatively few firms, each of which must take into account the reactions of its rivals to its own behavior (adapted from W.W. Norton and Company, 2003, *Glossary*. Available at <http://www.wwnorton.com/college/econ/stiglitz/gloss.htm>).

ONTOLOGY. An explicit formal specification of how to represent objects, concepts, and other entities that are assumed to exist in some area of interest, and the relationships among them.

OPEN ACCESS CONTENT *(as used in this report.)* Content openly available for others to access, use, and copy, and often to make derivative works, although some limited restrictions may apply. Typical restrictions may include preventing users from removing creator attribution from content, imposing identical license terms on any derived works, barring commercial use without permission, and liability limitations. We note that this definition does not necessarily conform to the use of the phrase "open access" in other contexts, including scientific publishing.

ORTHOIMAGE. A specially processed image prepared from an aerial photograph or a remotely sensed image that combines the accuracy of a traditional line map with the detail of an aerial image.

---

[1]See < http://www.fgdc.gov/nsdi/nsdi.html>.

OWNERSHIP OF GEOGRAPHIC DATA (*as used in this report*). With reference to a vendor or licensor, the owner is in possession of information that is not publicly known and holds the information as a trade secret. In the case of information to which copyright applies, the licensor is the owner of the copyright. With reference to a licensee, the licensee has possession of a copy of the information and has exclusive or nonexclusive rights to use and make the information available to others without restriction.

PUBLIC DOMAIN INFORMATION (*as used in this report*). Information that is not protected by patent, copyright, or any other legal right, and is accessible to the public without contractual restrictions on redistribution or use.

PURCHASE *(as used in this report).* A transaction or arrangement (usually a contract, in which there is an exchange of value) in which the purchaser of the geographic data (which may be contained in a geographic work) obtains *unlimited* rights to use, copy, and disseminate the geographic data.

RASTER DATA. Organization of a geographic space into cells, with each cell containing attribute measures (e.g., reflectivity, elevation, frequency, concentration). Each cell in the grid is of equal size and the cell density establishes the *spatial resolution* of the grid.

SECONDARY USERS. Those who are not the intended direct beneficiaries of the government data as defined by the mandates and missions of the agency but who nevertheless access government data and use it directly.

SERVICES *(as used in this report).* The processes of obtaining, processing, or providing geographic information (see GEOGRAPHIC SERVICES).

SHRINK-WRAP LICENSE (see also CLICK-WRAP LICENSE). A license setting forth the terms under which a vendor sells a right to use a product. The license is typically printed on the packaging containing the medium on which the data are delivered.

TERTIARY USERS. downstream users who do not directly acquire data from government but gain access through others who may merely pass it on or have made major changes to it.

TRANSACTION COSTS. Costs that include, but are not limited to, time spent on internal meetings, negotiations with vendors, review by lawyers, and the logistics of copying and distributing data to any employee who requests it.

UPLIFT RIGHTS. Conditions in a license that allow future purchases by specified parties under specified terms and conditions without the need to negotiate a new license.

VALUE CHAIN. A series of steps in which value is added to raw data through such actions as processing, analysis, and enhanced presentation.

VECTOR DATA. Representation of real-world themes such as addresses (points), road networks (lines), or land parcels (polygons). The location and shape of the polygon, line, or point is determined by the coordinate position (e.g., latitude/longitude) of its node(s).

WEB SERVICES. Self-contained, self-describing, modular applications that can be published, located, and invoked across the Web. Web services perform functions that can be anything from simple requests to complicated business processes. Once a Web service is deployed, other applications (and other Web services) can discover and invoke the deployed service (Source: Open GIS Consortium On-Line Glossary, at <http//:www.opengis.org>).

Appendix F

Acronyms and Abbreviations

ABA	American Bar Association
ACM	Association for Computer Machinery
AFGE	American Federation of Government Employees
AP	Associated Press
APA	Administrative Procedure Act
ARCIMS	Internet Map Server for the ESRI (see below) Arcinfo product
ASCAP	American Society of Composers, Authors and Publishers
B2B	Business-to-Business
B2G	Business-to-Government
BLM	Bureau of Land Management
BMI	Broadcast Music, Inc.
BTS	Bureau of Transportation Statistics
CFR	Code of Federal Regulations
CRADA	cooperative research and development agreement
CSC	Coastal Services Center (NOAA)
DAR	Defense Acquisition Regulations
DARPA	Defense Advanced Research Projects Agency
DEM	digital elevation model
DMCA	Digital Millennium Copyright Act of 1998
DoD	Department of Defense
DOE	Department of Energy
DOT	Department of Transportation

DQA	Data Quality Act
DRM	digital rights management
EIA	Energy Information Administration
EOSAT	Earth Observation Satellite Company
EPA	Environmental Protection Agency
ESRI	Environmental Systems Research Institute, Inc.
EU	European Union
FAA	Federal Aviation Administration
FAIR	Federal Activities Inventory Reform Act
FAR	Federal Acquisition Regulations
FDA	Food and Drug Administration
FDL	Federal Depository Libraries
FEMA	Federal Emergency Management Agency
FGDC	Federal Geographic Data Committee
FHA	Federal Highway Administration
FOIA	Freedom of Information Act
FRA	Federal Railroad Administration
GDT	Geographic Data Technologies, Inc.
GIS	Geographic Information System
GPO	U.S. Government Printing Office
GPS	Global Positioning System
GSA	General Services Administration
HUD	Department of Housing and Urban Development
INS	International News Service
ISO	International Organization for Standardization
MAF	Master Address File
MAPPS	Management Association for Private Photogrammetric Surveyors
MOU	memorandum of understanding
NASA	National Aeronautics and Space Administration
NCCUSL	National Conference of Commissioners on Uniform State Laws
NCLIS	National Commission on Libraries and Information Science
NEXTMap	name of product offered by Intermap Technologies, Inc.
NGA	National Geospatial-Intelligence Agency (formerly NIMA)
NGO	nongovernmental organization
NHTSA	National Highway Traffic Safety Administration
NIMA	National Imagery and Mapping Agency (now NGA)
NOAA	National Oceanic and Atmospheric Administration
NRC	National Research Council

NRCS	Natural Resources Conservation Service
NRO	National Reconnaissance Office
ODC	Open Data Consortium
OECD	Organization for Economic Co-operation and Development
OGC	Open GIS Consortium, Inc. [Open Geospatial Consortium as of September 2004]
OMB	Office of Management and Budget
ORBIMAGE	Orbital Image Corp.
QE2	*Queen Elizabeth II* (Ocean Liner)
RAM	Random access memory
SeaWiFS	Sea-viewing Wide Field-of-view Sensor
SPOT	Systeme Probatoire Pour l'Observation de la Terre
TIGER	Topologically Integrated Geographic Encoding and Referencing system
UCC	Uniformed Commercial Code
UCITA	Uniform Computer Information Transactions Act
UNESCO	United Nations Educational, Scientific and Cultural Organization
URISA	Urban and Regional Information Systems Association
USC	United States Code
USDA	U.S. Department of Agriculture
USGS	U.S. Geological Survey